Geology of the country around Great Dunmow

The landscape of this district is dominated by a gently undulating plateau of drift deposits, which were mainly laid down when the Anglian ice-sheet covered the entire district. The drift masks most of the solid rocks, which largely comprise the Upper Chalk, the Woolwich and Reading Beds, and the London Clay; patches of shelly Crag deposits are also present.

Complex drift sequences fill buried, relatively steep-sided channels along the alignment of the present Cam and Stort valleys. These channels were excavated during the glacial episode; parts of them persisted into the subsequent interglacial period as lakes, in which typical lacustrine deposits were deposited.

This memoir describes the results of the geological survey of the district and the accompanying drilling and trenching programmes; it also gives an account of the economic geology and water supply.

Frontispiece Cowlands Farm Gravel Pit, Stebbing. Clay enriched, reddened (rubified) horizon at top of face overlies well-bedded Kesgrave Sands and Gravels (A13555)

BRITISH GEOLOGICAL SURVEY

R D LAKE
and D WILSON

Geology of the country around Great Dunmow

Memoir for 1:50 000 geological sheet 222
(England and Wales)

CONTRIBUTORS

Stratigraphy
A Horton
D Millward
G Richardson
J A Zalasiewicz

Biostratigraphy
R Harland
M J Hughes

Hydrogeology
R A Monkhouse

LONDON: HMSO 1990

© NERC copyright 1990

First published 1990

ISBN 0 11 884473 3

Bibliographical reference

LAKE, R D, and WILSON, D. 1990. Geology of the country around Great Dunmow. *Memoir of the British Geological Survey*, Sheet 222 (England and Wales).

British Library Cataloguing in Publication Data

A CIP catalogue record of this book is available from the British Library

Authors

R D Lake, MA
British Geological Survey, Keyworth
D Wilson, BSc, PhD
British Geological Survey, Aberystwyth

Contributors
A Horton, BSc
J A Zalasiewicz, BSc, PhD
British Geological Survey, Keyworth
D Millward, BSc, PhD,
British Geological Survey, Newcastle
G Richardson and M J Hughes
formerly *British Geological Survey*

Other publications of the Survey dealing with this district and adjoining districts

BOOKS

British Regional Geology
London and the Thames Valley (3rd edition)

Memoirs
Huntingdon and Biggleswade (187, 204), 1965
Braintree (223), 1986
Epping (240), 1987
Chelmsford (241), 1985

Well Catalogue
Great Dunmow (222)
Braintree (223)

Mineral Assessment Reports
No. 46   North of Harlow, Essex TL41, 1979
No. 52   Hatfield Heath and Great Waltham, Essex TL51/61, 1980
No. 104 Stansted Mountfitchet, Essex TL52, 1981
No. 109 Great Dunmow, Essex TL62, 1982
No. 133 North-east of Thaxted, Essex TL63, 1982

MAPS

1:625 000

Solid geology (south sheet)
Quaternary geology (south sheet)
Aeromagnetic (south sheet)
Bouguer anomaly (south sheet)

1:125 000
Hydrogeological map of southern East Anglia

1:100 000
Hydrogeological map of the area between Cambridge and Maidenhead

1:50 000 or 1:63 360

Sheet 204 (Biggleswade), 1976
Sheet 205 (Saffron Walden), 1932
Sheet 206 (Sudbury), 1950
Sheet 223 (Braintree), 1982
Sheet 239 (Hertford), 1978
Sheet 240 (Epping), 1981
Sheet 241 (Chelmsford), 1975

Printed in the UK for HMSO
Dd 291126 C10   10/90

CONTENTS

One Introduction 1
Location and physical features 1
History of research 2
Geological history and structure 2

Two Pre-Tertiary formations 4
Palaeozoic 4
Cretaceous 4
  Gault and Upper Greensand 4
  Chalk 4
Details 5

Three Tertiary 7
Thanet Beds and Woolwich and Reading Beds (Palaeocene) 8
London Clay (Eocene) 8
Details 10

Four Plio-Pleistocene: Crag 14

Five Quaternary: Bramertonian to Anglian 16
Kesgrave Sands and Gravels 17
Glacial Sand and Gravel 18
Till (Boulder Clay) 18
Glacial Silt 19
Inter-relationships of the deposits 19
Details 19

Six Quaternary: Anglian glacial channels 28
Introduction 28
The Cam–Stort buried channel 28
  Morphology and fill of the channel complex 28
  Sedimentary relationships within the Cam–Stort buried channel 32
  Evolution of the Cam–Stort buried channel 33
Details 33

Seven Quaternary: Hoxnian to Flandrian 40
Lacustrine deposits 40
Head and Head Gravel 40
River Terrace Deposits 40
Alluvium: mere deposits 40
Alluvium: floodplain deposits 41
Calcareous tufa 41
Details 41

Eight Economic geology 42
Building materials 42
Sand and gravel 42
Hydrogeology and water supply 42

References 44

Appendices 47
1 Abstracts of selected borehole logs 47
2 1:10 000 maps 48
3 Open-file reports 49
4 Geological Survey photographs 50

Index 51

FIGURES

1 Physical features of the Great Dunmow district 1
2 Isobases and isopachytes of the Lower London Tertiary Group 7
3 Graphic log of Great Bardfield Borehole [6690 3091] 9
4 Generalised distribution of the Red Crag and its probable oxidised equivalents 14
5 Contours on the surface of the Cretaceous and Tertiary strata 15
6 Generalised distribution of the Kesgrave Sands and Gravels 17
7 Ugley sand quarry: section in north-east corner 19
8 Generalised distribution of the Glacial Sand and Gravel and Glacial Silt in the eastern part of the district 20
9a Plan of gravel pit at Cowlands Farm, Stebbing 26
9b Periglacial structures in face B of the gravel pit 26
10 Sketch section of north face of sand pit at Great Sampford 27
11 Morphological and lithological elements of the Cam–Stort buried channel complex beneath 'Plateau till' 29
12 Cross-sections of the Cam–Stort buried channel complex 31
13 Stages in the development of a buried channel 34

TABLES

1 Subdivisions of the Chalk 4
2 Tertiary stratigraphical nomenclature, as used in this memoir 8
3 Late Pliocene and Quaternary deposits of the Great Dunmow district 16
4 Boreholes proving the buried channel of the River Stort 32
5 Details of licensed water abstraction 43
6 Typical groundwater analyses from the Chalk 43

PLATES

Frontispiece Cowlands Farm Gravel Pit, Stebbing, Clay enriched, reddened (rubified) horizon at top of face overlies well-bedded Kesgrave Sands and Gravels
1 Pinchpools Chalk Pit. Fractured chalk with lenticular injections of silt following the flint courses 6
2 Newport Gravel Pit. Plan view of solution pipes on a scraped chalk surface 6
3 Elsenham Sand Pit. Pale grey, very chalky till overlies Kesgrave Sands and Gravels 23

NOTES

Throughout the memoir the word 'district' refers to the area covered by the 1:50 000 geological sheet 222 (Great Dunmow).

National Grid references are given in square brackets; all lie in 100 km square TL.

PREFACE

This memoir describes the geology of the district covered by the 1:50 000 Great Dunmow (222) New Series Sheet of the Geological Map of England and Wales. The original geological survey of Old Series Sheet 47, on the scale of one inch to one mile, was carried out by Messrs F J Bennett, W H Dalton and W H Penning, and published as separate solid and drift editions during the period 1881 to 1884.

The main part of the sheet was resurveyed by Drs D Millward, D Wilson and J A Zalasiewicz, and Messrs A Horton, R D Lake, P I Manning and G Richardson on the 1:10 000 scale in 1979–1984. The southern and eastern margins were mapped earlier by Dr B S P Moorlock and Messrs R A Ellison, M J Heath and S R Mills in 1975–1976 as part of the surveys of the adjoining Epping and Braintree districts. The work was carried out under the direction of Drs R A B Bazley, W A Read and R G Thurrell, District Geologists.

The memoir has been written largely by Mr Lake and Dr Wilson, who have incorporated information published in open-file reports compiled by Drs Millward and Zalasiewicz, Messrs Horton and Richardson, and themselves. Biostratigraphical contributions have been provided by Dr R Harland and Mr M J Hughes on dinoflagellate cysts and foraminifera respectively. The account of the hydrogeology was prepared by Mr R A Monkhouse. Photographs were taken by Mr C J Jeffery. Dr R W O'B Knox advised on Tertiary chronostratigraphy.

Thanks are due to the landowners, the local authorities and the Anglian Water Authority for their helpful co-operation.

F G Larminie, OBE
Director

British Geological Survey
Keyworth
Nottingham
NG12 5GG

1 March 1990

ONE

Introduction

LOCATION AND PHYSICAL FEATURES

This memoir describes the district covered by the Great Dunmow 1:50 000-scale Geological Sheet (222). The district, which lies some 40 km north-north-east of London, includes parts of Cambridgeshire, Essex and Hertfordshire. There is an extensive cover of glacial and postglacial drifts throughout the district, with till (boulder clay) as the most common glacial deposit. Beneath the drift, the district is underlain by Tertiary deposits in the south and east, and by the Upper Chalk elsewhere. The regional dip of the solid formations is gentle towards the south-east.

The drainage is directed mainly southwards by the rivers Ash, Stort, Roding, Chelmer and Pant, and their tributaries; a small area around Newport and Audley End is drained northwards by the headwaters of the River Cam. These streams dissect a broad plateau largely composed of boulder clay whose surface falls from 147 m near Nuthampstead in the north-east to 70 m near Felsted in the south-east (Figure 1). The district is dominantly agri-

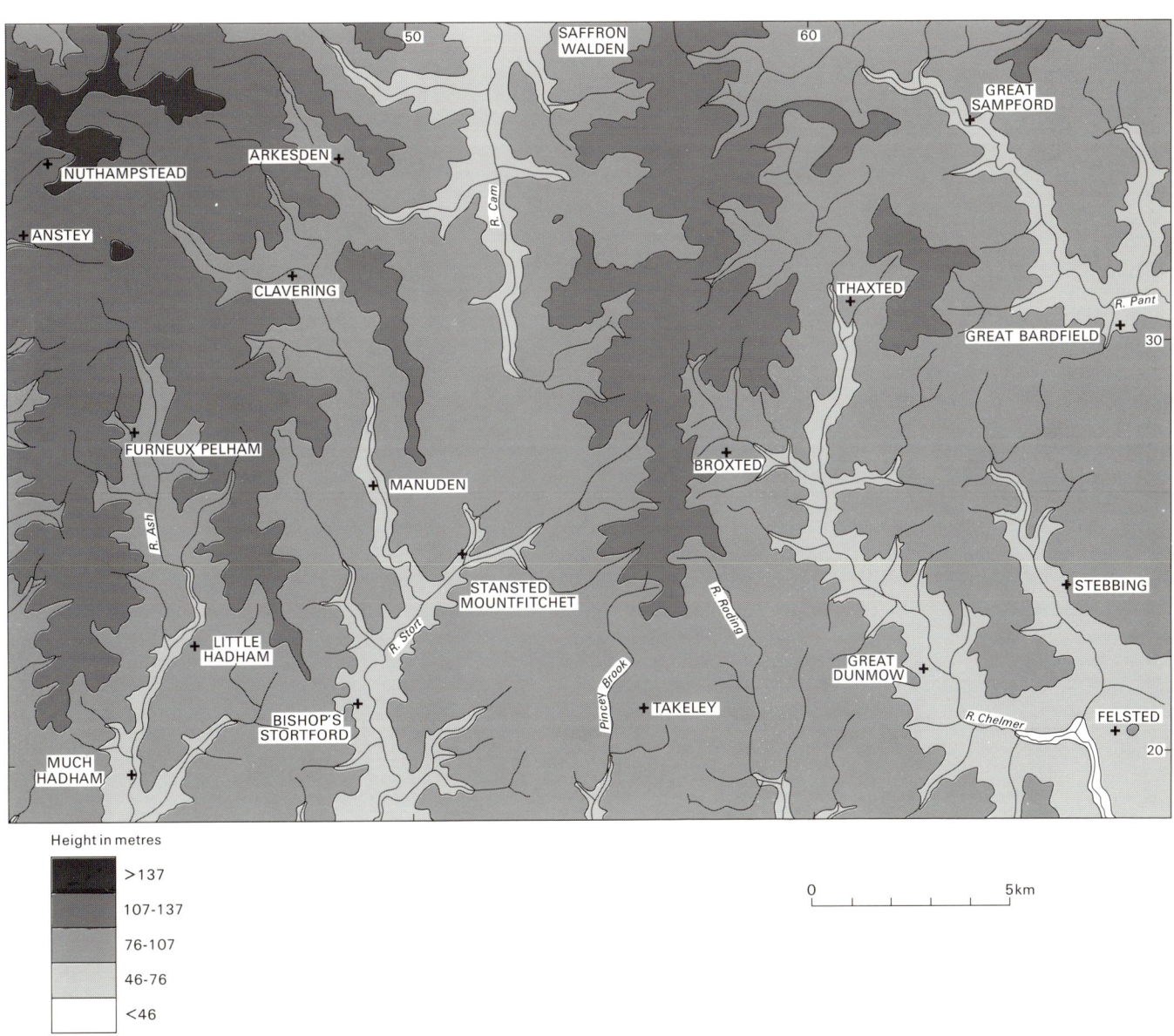

Figure 1 Physical features of the Great Dunmow district

cultural; arable farming is practised on all but the poorly drained heavy soils of some lower valley slopes and floors. The main centres of population are the expanding settlements of Bishop's Stortford and Stansted Mountfitchet, and market towns such as Great Dunmow and Thaxted; the outskirts of Saffron Walden encroach on the northern margin of the district near Audley End.

HISTORY OF RESEARCH

The Old Series geological one-inch drift and solid editions of Sheet 47 were published between 1881 and 1884. The descriptive account by Whitaker, Penning, Dalton and Bennett (1878) followed an earlier classification of the deposits by Prestwich (1847). The present survey is the first extensive geological mapping on the 1:10 000 scale of this district. Borehole and well records were included in the original memoir and were subsequently updated by Whitaker and Thresh (1916) and, more recently, in an inventory by Sayer and Harvey (1965). The hydrogeology of the district is portrayed on the 1:250 000 Hydrogeological Map of East Anglia (Sheet 2).

Regional studies of the Tertiary strata pertinent to this district have been carried out by Stamp (1921), Wrigley (1924, 1940) and Hester (1965). Curry (1958, 1965) synthesised the researches of contemporary workers in this field. Subsequently, King (1981) studied the London Clay in both the Hampshire and London basins, and Morton (1982) discussed the provenance of the Palaeogene sandstones of the Essex and London areas.

The Quaternary deposits of the district received little attention after an initial description of them by Whitaker et al. (1878). The deposits in the Cam valley were, however, described by Sparks and West (1965) and Baker (1976). Hey (1965, 1980) studied gravels to the west of Bishop's Stortford. Deposits providing resources of sand and gravel have been described in BGS mineral assessment reports (Hopson, 1979, 1981; Marks, 1980, 1982; Thomas, 1982). Regional syntheses which have a bearing on this district include those of Prestwich (1890), Solomon (1932), Mitchell et al. (1973), Rose and Allen (1977) and Mathers and Zalasiewicz (1988).

GEOLOGICAL HISTORY AND STRUCTURE

The oldest rocks proved in boreholes from this and surrounding districts are of late Silurian and Devonian age. They form the uppermost part of the London Platform, a basement block of Palaeozoic rocks which extends from southern central England eastwards to northern France and which has remained tectonically stable since Carboniferous times.

The earliest Mesozoic sedimentary rocks proved in the district are of the Upper Gault, deposited during the mid-Cretaceous marine transgression that extended across the London Platform. This episode of clay sedimentation was followed by a slight hiatus and subsequent deposition of the calcareous muds of the Chalk sea. Uplift and emergence, associated with mild deformation, terminated chalk deposition and initiated erosion of the highest Cretaceous strata.

Pulsed marine transgression across a slowly subsiding marginal basin, linked to continuing downwarp of the much deeper North Sea basin, occurred in early Tertiary times, with concomitant deposition of marine clastic sediments. The stratigraphy of the oldest Tertiary sediments in this district, the Thanet Beds, is still poorly understood; it is probable that considerable erosion of them occurred before the deposition and overlap of the more widely recognised Woolwich and Reading Beds. These latter sediments reflect a change from an open marine depositional environment, as represented by the pebbly 'Bottom Bed', to coastal deltaic swamp conditions, characterised by the distinctive red and grey mottled upper part of the Woolwich and Reading Beds, which indicates emergence and fluctuating water table conditions.

Subsequent deepening of the basin led to the deposition of the sandy marine basal beds of the London Clay, followed by a sequence of argillaceous sediments. The presence of thin ash bands in the lower part of the sequence, both in this and the Sudbury district to the north-east, indicates episodes of volcanic activity (Knox and Ellison, 1979). Younger Tertiary formations were probably once present in the Great Dunmow district but were removed by erosion following the Miocene earth movements. This episode of deformation created the broad syncline known as the London Basin.

The remains of a broad peneplain, cutting across the Cretaceous and Tertiary strata, and falling from about 100 m in the west to 65 m in the east, may be present beneath the drift deposits of the plateau (Figure 5). Deposits which overlie the surface in the east suggest that it formed during one or more transgressions which spanned the Pliocene to early Pleistocene interval. The peneplain extends eastwards to the coast at Walton on the Naze where it lies at 15 m above OD; its inclination is probably related to the gentle tectonic downwarp of the Southern North Sea Basin.

The Crag deposits comprise shelly sands containing cold climate faunas that initiated a time of climatic fluctuations which ranged from warm temperate to glacial, and which have continued until the present day.

Overlying the Crag, and probably containing sands derived from it, is a suite of drift sediments, known collectively as the Kesgrave Sands and Gravels (Rose and Allen, 1976). The pebbly beds of this suite were probably deposited by a braided river, a 'proto-Thames', that flowed on a more northerly course than the present river, from near Reading, through St Albans to Bishop's Stortford, and thence eastwards across Essex. The sandy beds may relate to an earlier marine incursion.

The Anglian ice-sheet advanced in a southerly direction across the district some 500 000 years ago; the earliest incursion of ice may have occupied the Cam valley. The complex bedrock topography in this valley and southwards into the Stort valley indicates both deep subglacial erosion to form buried channels and, perhaps, the formation of spillways linking these two valleys at that time. When the ice-sheet wasted, a chalky clay lodgement till (boulder clay) was left as a blanket over much of the district. Other glacial sediments such as Glacial Sand and Gravel and Glacial Silt are more localised in occurrence, being particularly common in the buried channels.

As the climate ameliorated in the Hoxnian interglacial stage, small lakes were formed in hollows in the valley systems such as that near Quendon [524 303], in which tufaceous clays are preserved. Elsewhere, small lakes may have formed in kettle holes on the plateau surface.

There are few deposits in this district which record the later parts of the Quaternary. Up to two levels of River Terrace Deposits, representing former floodplain deposits, have been recognised in the Ash, Stort, Chelmer and Pant valleys.

Solifluction deposits (Head), which accumulated by downslope movement of material under periglacial climatic conditions, occur in many of the valley systems. Locally, some of the minor tributary valleys of the Cam, such as that near Bonhunt [512 334], were impounded by solifluction flows to create temporary meres in late Devensian to Flandrian times. During the Flandrian, and continuing until the present day, argillaceous alluvial sediments have been deposited along the floodplains of the major river valleys.

TWO
Pre-Tertiary formations

PALAEOZOIC

Only one borehole in the district has encountered Palaeozoic strata. The Little Chishill No. 1 Well [4528 3637], sunk by the Superior Oil Company in 1964, encountered Devonian sedimentary rocks at 143 m below OD. About 10.6 m of these strata were cored, comprising greenish grey laminated mudstones, silty striped mudstones and siltstones. The siltstones in the lower beds are reddish brown. Thin sparry limestones, some of which are thinly bedded with silty partings, also occur. Intraformational pebbles of limestone are present at intervals. The brachiopod and bivalve fauna indicates a shallow marine environment and a Frasnian or early Famennian age (Butler, 1981).

Two other deep boreholes were sunk close to the district boundary. One, at Ware [3531 1398], encountered shelly limestones of Wenlock age at 206 m below OD. The other, at Saffron Walden [5386 3840], is reputed to have terminated in Palaeozoic rocks at 145 m below OD (Whitaker et al., 1878), but the evidence of this and for the presence of Upper Jurassic sedimentary rocks is uncertain (White, 1932). Geophysical evidence suggests that much of the Palaeozoic basement of Essex comprises Upper Devonian sedimentary rocks (Allsop and Smith, 1988).

CRETACEOUS

Gault and Upper Greensand

Cretaceous strata probably rest directly on Palaeozoic rocks throughout much of the district. In Little Chishill No. 1 Well, 61.6 m of Gault rest unconformably on the Devonian strata at 280.0 m depth. The Gault sedimentary rocks which were cored between the depths of 255.72 and 267.33 m, comprise medium grey silty mudstones with a thin micritic limestone, pale phosphatic nodules and scattered fossils. In the Ware Borehole, 50.7 m of Gault and 12.2 m of Upper Greensand were recorded. One interpretation of the sequence in the Saffron Walden Borehole suggests that 47.2 m of Gault overlie 15.2 m of sandy beds (Carstone or Lower Greensand), which rests on Palaeozoic strata at 197.2 m depth.

It is probable that only the Upper Gault is present in this district; Beds 13 to 15 of the lithological subdivision and zonation of the Gault of East Anglia by Gallois and Morter (1982) are represented in the cores from Little Chishill No. 1 Well.

Chalk

The Chalk comprises mainly white, fine-grained, microporous limestones totalling up to about 220 m in thickness.

The Lower, Middle and Upper Chalk are the main divisions generally recognised in south-eastern England. The divisions, stages and zones of the Chalk likely to be present in this and adjoining districts are shown in Table 1. Only the Upper Chalk is present at outcrop in the district and, although the Middle Chalk is thought to occur beneath the drift of the Cam valley to the north of Newport, its extent is uncertain.

Table 1 Subdivisions of the Chalk

Stratigraphical division	Zone	Stage
Upper Chalk	*Marsupites testudinarius*	Santonian
	Uintacrinus socialis	-----
	Micraster coranguinum	Coniacian
	Micraster cortestudinarium	-----
	Sternotaxis planus	
	Terebratulina lata	
Middle Chalk	*Mytiloides labiatus*	Turonian
	Neocardioceras juddii	
	Metoicoceras geslinianum	
	Calycoceras guerangeri	
Lower Chalk	*Acanthoceras jukesbrownei*	Cenomanian
	Acanthoceras rhotomagense	
	Mantelliceras dixoni	
	Mantelliceras mantelli	

The lithostratigraphy of the Lower, Middle and Upper Chalk has been described in nearby districts (White, 1932; Worssam and Taylor, 1969). The Lower Chalk comprises pale grey marly chalks and marls with the distinctive Glauconitic Marl at its base. The Middle Chalk is off-white and contains bands of nodular chalk with few scattered flints in the upper part; at its base the nodular Melbourn Rock forms a distinctive marker. The Upper Chalk is white, with abundant courses of nodular flints at most levels; the nodular Chalk Rock is present at its base and, about 10 m above, there is a mineralised hardground, the Top Rock.

Although the Melbourn Rock and Chalk Rock are readily recognised at outcrop elsewhere, they and other hard bands are not usually identified in the lithological logs of boreholes and wells sunk for water. Consequently, the older records are difficult to classify. With the development of geophysical logging, the combination of resistivity and gamma logs has provided an important tool for borehole correlation (Murray, 1986).

Boreholes at Little Chishill [4213 3941; 4528 3637] proved the following composite sequence (based mainly on geophysical logs):

	Thickness m
Upper Chalk	44.4
Middle Chalk	81.1
Lower Chalk (44.5 m)	
Grey Chalk (including Plenus Marls)	26.7
Totternhoe Stone; gritty-textured hard chalk	2.5
Chalk Marl? and Glauconitic Marl	15.3

Almost the entire thickness of the Lower Chalk may have been penetrated (48.7 m) in a borehole [4402 3775] at Chrishall. Boreholes to the north-east at Strethall [4919 4143] and Saffron Walden [5386 3840], north of the district, appear to have penetrated 54.5 and 54.9 m of Lower Chalk respectively.

There are some discrepancies between the lithological logs, which lack descriptive detail, and the geophysical logs of some boreholes in the Middle Chalk. In consequence, estimates of the thickness of these beds vary between 65 m and 111 m. The more reliable records of boreholes at Chrishall [4402 3775] indicate thicknesses of 73.9 m or 77.3 m; another at Shaftenhoe End [4143 3766] indicates a thickness of 74.7 m. The Middle Chalk may be thinner to the east, 64.9 m having been recorded at Saffron Walden [5386 3840]. If a consistent regional dip of 1 in 150 is assumed for the Chalk, it is estimated that up to 90 m of Upper Chalk may be present in the western part of the district.

The highest Chalk preserved in this district probably belongs to the *Micraster coranguinum* Zone (Curry, 1965), in common with much of the Saffron Walden district (White, 1932).

DETAILS

A disused quarry [4397 2311], 400 m north of Little Hadham, in the north-east corner showed:

	Thickness m
Soil	0.5
HEAD GRAVEL	
Gravel and clayey sand	1.0
?KESGRAVE SANDS AND GRAVELS	
Sands, medium-grained, cross-bedded, brown (filling shallow channels in the chalk surface)	0 to 2.0
CHALK	
Chalk, white, with several horizons containing abundant black, irregular, nodular flints	8.0

In an old chalk quarry 450 m east of Albury Hall [4323 2553] about 2.5 m of soft white chalk with some black irregular flints are exposed.

An exposure in the bed of the River Ash [4375 2557], adjacent to an exposure of Head Gravel, shows soft reconstituted chalk with pebbles incorporated to a depth of 0.3 m or more below its upper surface. Another weathering feature of the Chalk is seen in a ditched stream [4036 2685] 800 m south-west of Bozen Green, where soft white chalk about 1.5 m thick is exposed below about 1.0 m of gravelly Head. Solution hollows can be seen in the chalk surface to a depth of 1.5 m or more. These are lined with layers of yellow to orange sharp sand and red to brown clay, which drape around the hollows. Also, within the upper part of the chalk, there are lenticular horizontal cavities up to 3.0 m in diameter filled with similar deposits. The disposition of these solution features appears to be controlled by the dominant vertical and horizontal jointing.

At Pinchpools, a pit [4920 2758] exposed 4 m of soft to firm white chalk with nodular flint horizons beneath up to 1 m of chalky wash. A prominent tabular flint band, 2 cm thick, was noted 3 m below ground surface. The chalk showed intense vertical fracturing and local subhorizontal shears; the latter tended to follow the flint courses and contained lenticular cryostatic injections of silt or silty sand with clay laths (Plate 1). The tabular flint course showed local kink-flexures. Silty chalk-breccia filled some of the more prominent joints.

The only large exposure of chalk is in a quarry [525 331] near Newport railway station where 20 m of fractured white Upper Chalk is exposed; a persistent layer of tabular flint occurs about 6 m below the top of the face. Several solution pipes were visible at the top of the old quarry face and many more were exposed on the scraped chalk surface to the east (Plate 2); they were typically 1 to 2 m wide and 2 to 3 m deep, lined with *remanié* dark brown sandy clay and infilled with reworked Kesgrave Sands and Gravels.

Up to 4 m of vertically fractured white chalk with nodular flint layers are exposed beneath the working floor of a former gravel pit [515 265] near Alsa Street. The chalk is penetrated by solution pipes up to 2 m wide and 3 m deep, lined with reworked glauconitic Tertiary material and filled with brown sandy clays and clayey sands containing gravelly lenses.

The chalk pit [5170 2505] near Stansted Castle formerly exposed some 8 m of Upper Chalk cut by a small fault (Whitaker et al., 1878, fig. 7).

6 CHAPTER TWO PRE-TERTIARY FORMATIONS

Plate 1
Pinchpools Chalk Pit. Fractured chalk with lenticular injections of silt following the flint courses (A 13553)

Plate 2
Newport Gravel Pit. Plan view of solution pipes on a scraped chalk surface (A 13567)

THREE

Tertiary

The Tertiary deposits comprise a sequence of sands and clays that lies unconformably upon the Upper Chalk and beneath the widespread cover of drift deposits. These beds were laid down in a slowly subsiding sedimentary basin with its centre in the region of the Thames Estuary, which was connected to the much larger Southern North Sea Basin. The position of the basin margin is uncertain because post-Eocene erosion removed the Tertiary strata in the northern part of this district. Contours on the basal surface of the Tertiary deposits are shown in Figure 2, which also depicts the isopachytes of the combined Thanet Beds and Woolwich and Reading Beds (Lower London Tertiary Group).

The Tertiary sediments fall within the Palaeocene and Eocene series but the precise positions of the series boundary and stage boundaries (Table 2) are uncertain. This stems partly from the scarcity of biostratigraphically significant fossils in the UK sequences and partly from the lack of international agreement on the definition of the series and stage boundaries. There is, however, a general consensus among biostratigraphers working on the world-wide deep-sea record that the Palaeocene – Eocene boundary is best placed at the base of calcareous nannoplankton zone NP10. Since this is the definition adopted on most recent Palaeogene time-scales (e.g. Berggren, Kent and Flynn, 1985), it is provisionally adopted here. Direct identification of the NP9 – NP10 zonal boundary is not, however, possible because of the absence of calcareous nannoplankton over the relevant interval. Nevertheless, indirect correlation with the biozonal scheme is made possible by the existence in the basal part of the London Clay of East Anglia and Essex of a series of volcanic ash layers that can be recognised over much of north-west Europe and that can be dated positively as of earliest NP10 zone age by their occurrence in nannofossil oozes in the eastern Atlantic (Knox, 1984). The ash layers are now known to be restricted to the London Clay, and since a hiatus exists throughout southern England between the London Clay and the

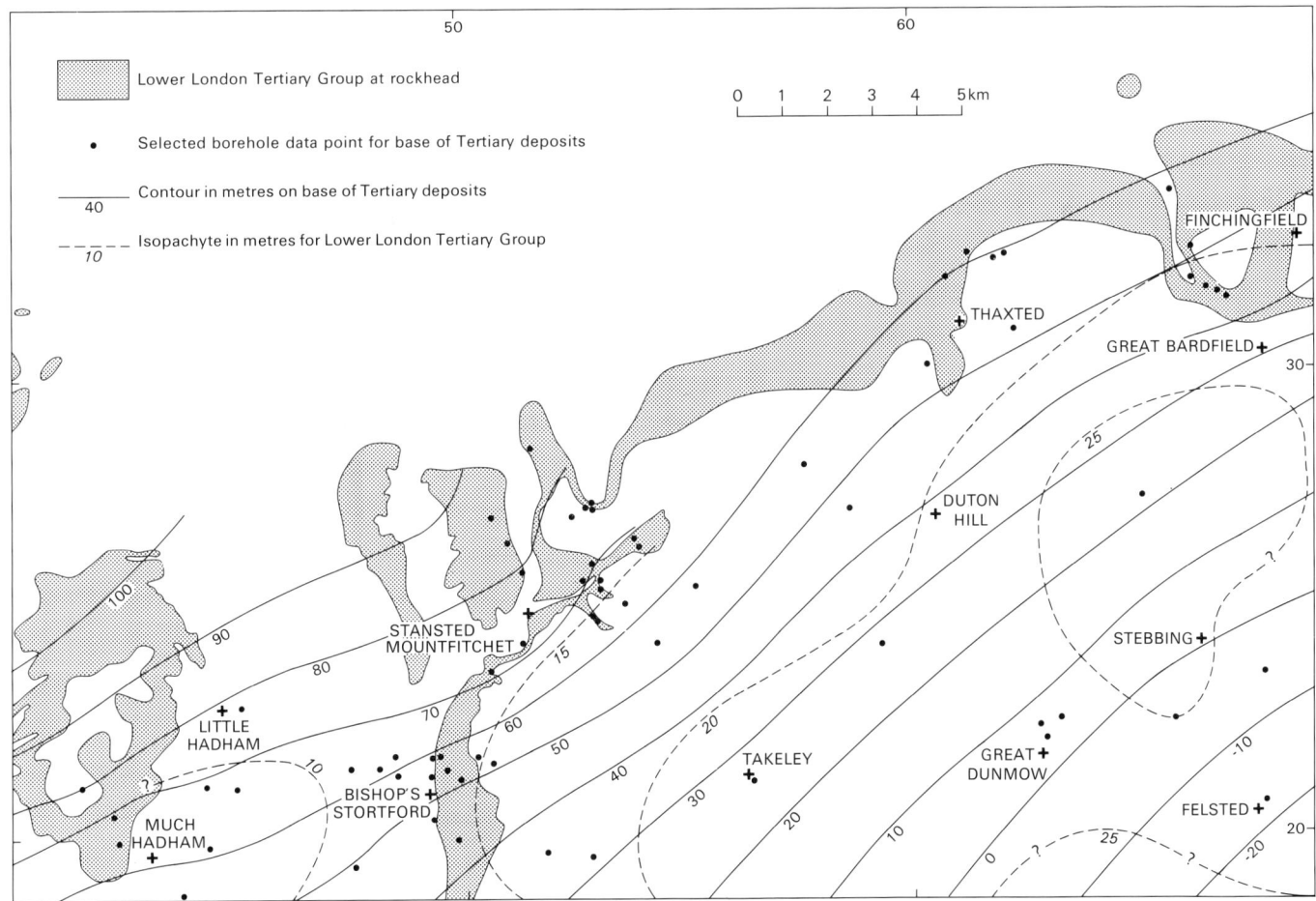

Figure 2 Isobases and isopachytes of the Lower London Tertiary Group

Table 2 Tertiary stratigraphical nomenclature, as used in this memoir

System	Series	Stage	Group	Formation
Tertiary	Eocene	Ypresian		London Clay
	Palaeocene	'Sparnacian'	Lower London Tertiary Group	Woolwich and Reading Beds
		Thanetian		Thanet Beds

underlying Woolwich and Reading Beds, it seems reasonable to draw the NP9–NP10 boundary and hence the Palaeocene–Eocene boundary at the base of the London Clay. By inference the base of the Ypresian may also be taken at the base of the London Clay.

THANET BEDS AND WOOLWICH AND READING BEDS (PALAEOCENE)

In this account the lithostratigraphical subdivision of the Lower London Tertiary Group follows that of Whitaker (1866; after Prestwich, 1852). Of the three divisions, the Thanet Beds, the Woolwich and Reading Beds and the Oldhaven Beds, only the first two have been recognised in this district. These beds are patchily exposed between Much Hadham [43 19] and Ugley Green [52 29]. Elsewhere, interpretation of the sequence is hampered by the poor descriptions of many of the boreholes which penetrated the Tertiary deposits. However, it is thought that throughout the outcrop of the Lower London Tertiary Group the Thanet Beds have been largely overlapped by the Woolwich and Reading Beds. A Thanetian microflora has, however, been recovered from green and brown mottled fine-grained sands exposed in an excavation for the M11 motorway [about 530 255] near Stansted Mountfitchet (Harland, BGS internal report No. PDL 78/228). Up to 10 m of Thanet Beds have been recorded in the subcrop near Great Dunmow and in Great Bardfield Borehole (Figure 3).

The Thanet Beds consist of brown and grey, glauconitic, silty, fine-grained sands of a uniform appearance due to the homogenising effects of extensive bioturbation. The Bullhead Bed, at the base, consists of about 1 m of large, dark greenish grey-coated nodular flints in a glauconitic sandy clay matrix. Mineralogical studies of boreholes in the adjacent Braintree district (Ellison and Lake, 1986) have shown the significant presence of volcanogenic materials, including pyroclasts and zeolites, particularly near the base of the Thanet Beds.

The Woolwich and Reading Beds comprise two main lithologies. The lower beds are dominated by greenish grey, fine- to medium-grained, locally very glauconitic sands; the upper and thicker part generally comprises red, grey, green and brown mottled clay. These two types broadly correspond to the 'Bottom Bed' facies and 'Reading type' (or facies) of Hester (1965) respectively. At outcrop the basal, pebbly part of the Bottom Bed is similar to the Bullhead Bed, but the flints are smaller (less than 5 cm diameter) and include both fresh, black, rounded (about 2 cm diameter) and bleached nodular forms. It is thought that this unit incorporates recycled material from the Thanet Beds (Hester, 1965, p.121). Borehole data rarely indicate flint pebbles at the base of the Woolwich and Reading Beds, and the boundary is generally taken at the level where both the grain-size and glauconite content increase. The glauconitic and pebbly Bottom Bed is widely accepted to be the product of a short-lived marine transgression. The maximum thickness of the Woolwich and Reading Beds is thought to be about 17 m near Great Dunmow. The clay mineralogy of the two main lithologies has been shown to be distinctive (Ellison and Lake, 1986). The lower beds are smectitic and may have a volcanogenic component, whereas illite is dominant in the upper beds, accompanied by significant amounts of kaolinite.

From a study of the Woolwich and Reading Beds to the south-west of this district, Bateman and Moffat (1987) concluded that, there, the lower sands, above the Bottom Bed, and the upper mottled clays are texturally distinct and do not comprise a simple fining-upwards cycle. Their contact may demarcate different environmental parameters, namely lagoonal and deltaic sedimentary regimes respectively.

Near the margins of the outcrop of the Woolwich and Reading Beds, the surface of the Chalk is commonly irregularly pock-marked by solution pipes. These structures, such as those in the exposure [515 265] near Alsa Street, are up to 2 m wide and 3 m deep and are filled with reworked Tertiary and drift deposits. Isolated pockets of Woolwich and Reading Beds occur in similar situations as far north as Widdington, up to 3 km north of the main crop, and near Great Hormead; the small outliers depicted on the published map may also have been disturbed by the effects of solution. There are few permanent exposures in the Woolwich and Reading Beds.

LONDON CLAY (EOCENE)

The London Clay is the youngest Tertiary formation to be preserved in the district. It commonly crops out in the floors of the valleys in the south and east of the district, where fluvial erosion has cut down through the blanket of drift deposits. This formation comprises mainly olive-brown to grey micaceous silty clays which weather chocolate-brown, due to oxidation of the contained pyrite; siltstones and septarian nodules are present locally. At the base, there are dark olive-grey to greenish grey, silty, fine-grained sands, with small rounded black flint pebbles up to 1 cm in diameter. These sands rest disconformably on the Woolwich and Reading Beds and are 5.6 m thick in Great Bardfield Borehole (see Figure 3); boreholes elsewhere generally indicate thicknesses of between 0.5 and 5 m.

Farther south in Essex, the London Clay has been divided into six lithological units (labelled LA to LF in ascending order) (Bristow, 1985; Lake et al., 1986) and into biofacies units by King (1970; 1981). The dinoflagellate cyst assemblages recognised in the basal London Clay appear to be relatively free from facies-control and demonstrate that the London Clay transgression proceeded westwards, probably in two main stages (Knox, Harland and King, 1983).

Figure 3 Graphic log of Great Bardfield Borehole [6690 3091]

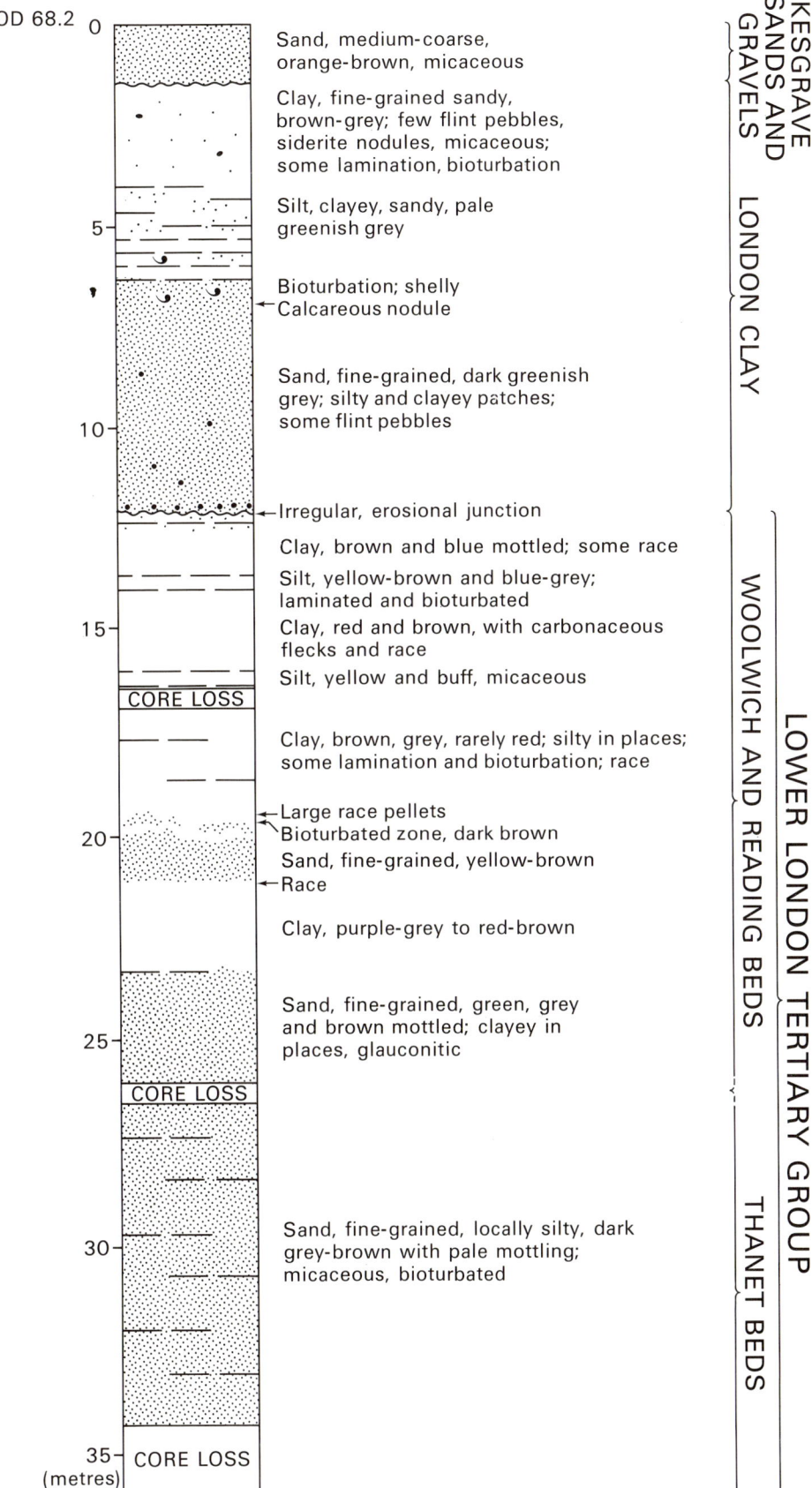

There is little borehole core material available for inspection from the district, but it is thought that the basal sandy beds correspond to units LA and LB and that much of the clay of the remaining outcrop falls within unit LC. Up to 65 m are present in the district; in the Epping district to the south, the full thickness of the London Clay is between 115 and 132 m.

Tuffaceous (volcanic ash) bands have been observed in the lowest beds of the London Clay in this and the adjoining Sudbury district (Knox and Ellison, 1979). For example, the brownish grey sandy clays, which were exposed during the construction of the M11 motorway near Elsenham [533 255], contained pale buff tuffaceous partings 2 to 3.5 m above the base of the formation.

In Great Bardfield Borehole [6690 3091] Mr M J Hughes, formerly of the Biostratigraphy Research Group, reports that the following species are present in the London Clay part of the sequence: Foraminifera: *Chiloguembelina wilcoxensis* (Cushman & Ponton), *Epistominella* cf. *vitrea* (Parker), *Glandulina* sp.1, *Glandulina* sp. 2, *Globulina ineaqualis* Reuss, *Glomospira?* sp., *Gyroidina* sp., *Haplophragmoides* sp., *Nonion* cf. *graniferum* (Terquem), *N.* cf. *laeve* (d'Orbigny), *Nonionella* sp., *Protelphidium* cf. *hofkeri* Haynes, *P.* sp. 3 Murray & Wright 1974 and *Pseudohastigerina wilcoxensis* (Cushman & Ponton); Ostracods: *Clitherocytheridea* sp. 1, *Cytheretta* aff. *scrobiculoplicata* (Jones) and *Cytheropteron brimptoni* Bowen. The assemblage is similar to, though less diverse than that recovered from basal London Clay elsewhere in the northeast of the London Basin. A palaeoenvironment of very shallow water with restricted marine connections is thought probable.

The Woolwich and Reading Beds yielded only reworked Upper Cretaceous specimens whilst the Thanet Beds were completely barren of microfauna.

DETAILS

In a temporary exposure [4045 2088] at Standon Lodge the following section was recorded:

	Thickness m
HEAD	
Clay, silty, brown and yellow, with irregular patches of gravel	1.2
WOOLWICH AND READING BEDS	
Sand, very fine-grained, silty, yellow and grey mottled	0.6
Clay, very silty, red and grey mottled with grey bands; some beds of silty fine-grained sand	0.9

In the Pumping Station Borehole, Bromley Road, Much Hadham [4160 2104], the following sequence was described:

	Thickness m	Depth m
DRIFT	2.51	2.51
LONDON CLAY		
Clay, brown and blue, sandy and pebbly at base	15.02	17.53
WOOLWICH AND READING BEDS		
Clay, mottled	5.64	—
Clay, dark, tough	0.61	—
Clay, sandy, mottled	1.37	—
Clay, sandy and pebbles	0.46	—
Clay, sandy, mottled	1.37	26.98
UPPER CHALK	55.93	82.91

The section in an old brickyard east of Hadham Ford [4385 2165] was as follows:

	Thickness m
HEAD	
Clay, sandy, brown, with pebbles	0.65
LONDON CLAY	
Clay, silty, buff to brown, with thin sandy bands	1.50
Limestone, sandy, buff, with abundant shell fragments	0.25
Clay, buff to brown, with thin buff to yellow sandy bands	0.75

Subsequently, this exposure was obscured by a mass of boulder clay which slipped down the old quarry face. A section in this brickyard was recorded in 1869 (Whitaker et al., 1878) as follows:

	Thickness m
LONDON CLAY	
Brown clay, rather sandy, especially towards the bottom; with patches and darker lumps of tenacious clay, some pieces of septaria and lumps of 'race' (secondary calcareous concretions), impressions of shells and an occasional small pebble or piece of ironstone. Near the bottom two or three thin layers of dark clay	4.27
Yellowish, finely bedded sand	0.30
Yellowish, finely bedded sand full of shells, too friable for preservation, with rare pebbles: in places a thin layer of clay between the two beds of sand	0.30
'?READING BEDS'	
Very fine white sand, finely bedded, with some dark grains, and here and there small lumps of dark clay	2.44

At Churchend Farm, east of Little Hadham, a temporary exposure [4481 2288] in the footings of a barn gave the following section:

	Thickness m
WOOLWICH AND READING BEDS	
Sand, medium-grained, yellow, with 'sharp' angular grains, and dark green grains of glauconite (?reworked)	0.65
Sand, fine-grained, silty, red, with clay bands	0.33
Sand, fine-grained, silty, green, glauconitic	0.25

A roadside section [4843 2741] east of Maggots End exposed 1 m of glauconitic silts (basal Woolwich and Reading Beds) beneath a 'draped' wash of slightly pebbly medium-grained sands. Red sandy clay was noted in the base of the section nearby [4845 2741].

A water well [4512 2278] at Hadham Hall proved the sequence:

	Thickness m	Depth m
TILL		
Clay, brown, with fragments of chalk and worn shells	3.7	3.7

LONDON CLAY

Clay, blue-grey	4.0	—
Clay, dark greenish grey	1.8	—
Loam, brown, sandy	1.8	—
Clay, dark grey, with traces of lignite	0.6	—
Loam, brown, sandy	0.6	—
Loam, grey, with iron-pyrites after wood	1.5	14.0

WOOLWICH AND READING BEDS

Clay, stiff, greenish brown	0.3	—
Loam, fine, dark grey	0.5	—
Loam, pale grey and brown, slightly mottled	1.1	—
Clay, very tough, hard, red	4.6	—
Sand, very fine, loose, grey	1.5	—
Loam, sandy, green and red mottled	1.8	—
Sand, compact, grey	3.0	—
Flints, small, rounded, green-coated, passing down into large angular green-coated flints	0.3	27.1

UPPER CHALK

Chalk with flints; broken and rubbly at top, more compact below; hard beds in lower parts	56.7	83.8

A trial borehole [4654 2192] near Hadham Park proved the sandy basal beds of the London Clay:

	Thickness m	Depth m
HEAD	1.7	1.7
LONDON CLAY		
Clay, firm, silty, sandy, brown and grey mottled	3.2	4.9
Sand, silty, clayey, grey, with sporadic fossils and lenses of clay; band of shelly sandstone from 5.40 to 5.45 m	4.3	9.2
WOOLWICH AND READING BEDS		
Clay, silty, sandy, stiff, brown mottled, fissured, with blue veins and calcareous nodules	6.6	15.8
Sand, clayey, silty, multicoloured, mottled	4.2	20.0

A nearby borehole [4653 2188] proved the base of the Tertiary sequence at 69.6 m above OD, and thus the thickness of Woolwich and Reading Beds in this area is 12.8 m.

A borehole [4776 2286] near Wickham Hall proved a pebble bed at the base of the London Clay:

	Thickness m	Depth m
LONDON CLAY		
Clay, silty, sandy, firm, pale brown and grey mottled	4.7	4.7
Clay, silty, sandy, firm, grey, slightly fissured, becoming more sandy and containing sporadic fossils in lower levels	6.5	11.2
Clay, silty, sandy, grey, with many pebbles	0.1	11.3
WOOLWICH AND READING BEDS		
Clay, silty, stiff, fissured, brown, with blue and sporadic red mottling; scattered calcareous nodules	5.5	16.8
Sand, pinkish brown, with a few thin clay lenses	0.2	17.0

A section [4804 2202] adjacent to Cricketfield Lane, Bishop's Stortford, exposed the basal beds of the London Clay; they comprised 2 m of brown sandy clays and clayey sands overlying 1 m of very soft, water-saturated, fine-grained sands with soft (?decalcified) siltstone nodules. Mr Doyle of Hadham School reported that dark clay with flint pebbles overlain by fossiliferous clay was observed in a nearby trench opposite Grailands [4796 2193].

Three adjacent brickyards in Rye Street, Bishop's Stortford, formerly exposed the Woolwich and Reading Beds (Whitaker et al., 1878, pp. 22–23). It is not clear from the description whether all the material observed was in situ, and for brevity only interpretative abridged sections are given here:

	Thickness m
1 Mr Cornwell's brickyard [4865 2213]	
Sands and pebbly loams irregularly interbedded and cross-bedded: the pebbles included a few flints and 'many nodules of bright red ironstone in all stages of decomposition'. (Probably solifluicted materials derived in part from the basal London Clay)	1.5 +
'Grey sandy loam with a thin bed of clay about the middle, above which it is mottled with deep red patches': clayey at base	5.2
'Brown clay, with green-coated flints and pebbles'	0.3
Upper Chalk	—
2 Mr Glasscock's brickyard [4860 2204]	
'Brown London Clay'	1.5
'Mauve sand', in one place only	up to 0.6
'Grey sandy loam, the upper part mottled'	—
3 Mr Dickenson's brickyard [4860 2195]	
'Brown London Clay'	3.0
'Very fine light mauve sand'	0 to 1.0
'Grey sandy loam' (as in section 1) with reputed 'impressions of shells'	4.6
'Brown clay'	1.2
'Brown clay with green-coated flints and pebbles'	0.3 to 0.5
Upper Chalk	—

It seems unlikely that any of these pits exposed London Clay *sensu stricto*. They now appear to have been regraded.

Another brickyard 'just south-east of the Nonconformist School' [Bishop's Stortford College] [4812 2142] was described by Whitaker et al. (1878, p. 23) as:

	Thickness m
Soil and wash	0.6
LONDON CLAY	
Clay, brown and tenacious, passing down into the bed below	0.6
Whitish rather clayey sand, passing down to	0.9
Whitish clayey sand. At one part near the base [0.05 m] of very dark sand, with race and an occasional flint pebble	2.7
WOOLWICH AND READING BEDS	
Green, finely bedded sand	0.6
Whitish clayey sand	2.1
Brown clayey sand	0.6
Dark green clayey sand	1.2
Hidden by slips	0.9?
UPPER CHALK	—

The thickness of obscured material in this section is probably much underestimated.

The following section including the basal glauconitic pebble bed was recorded in the Hollow Lane Gravel Pit [5318 3066] near Widdington:

	Thickness m
DRIFT (see p.22)	10.0
WOOLWICH AND READING BEDS	
Silts, clayey, glauconitic, bluish green, locally reddish brown mottled, especially at base; layer of large glauconite-stained nodular flints at base	0.6
UPPER CHALK	1.4

The basal bed is preserved in hollows in an irregular Chalk surface in the south-east corner of the pit.

The Ugley Sand Quarry revealed the following composite section, based on the lowest part of an exposure [5206 2792] and a trial pit beneath (Figure 7):

	Thickness m
WOOLWICH AND READING BEDS	
Clay, sandy, silty, glauconitic, buff to olive-grey, soft, structureless; thin layer of shell detritus at 2 m; isolated pods of fine ochreous sand and glauconitic clay pellets. (?destructured)	2.9
Clay, pale grey, sheared (ie with ?listric surfaces); passing down to less compact bluish pale grey clayey silt; passing down to red, purple and bluish grey intensely bioturbated compact clay with listric surfaces	0.1

The basal glauconitic pebble bed of the Woolwich and Reading Beds was noted in scrapings on the floor of Ugley Sand Quarry at two localities [5173 2788 and 5177 2763] to the west. The pebbles comprised small, rounded (?abraded) nodular flints.

The following possibly incomplete composite sequence is derived from boreholes for the M11 near Elsenham:

	Thickness m
WOOLWICH AND READING BEDS	
Clay, very sandy and silty, grey, locally laminated (possibly basal London Clay in part)	2.1
Clay, silty, brown to grey, with calcareous concretions	5.2 to 6.7
Clay, sandy and clayey sand, multicoloured	2.4 to 4.3
Clay, silty, grey, laminated	0 to 0.9
Sand, fine-grained, compact and grey silts	1.8 to 5.6
UPPER CHALK	—

Lenticular medium-grained sand bodies are present locally above the multicoloured sequence. The record of the Memorial Well [5365 2634] at Elsenham suggests, however, that the mottled beds may be patchily distributed because multicoloured clays were only thinly represented in the 14.5 m of the complete formation.

The basal Tertiary beds, which rest on Chalk in the pit [5170 2505] near Stansted Castle, were described by Penning (in Whitaker et al., 1878, p.24) as:

	Thickness m
Green sand, mottled red	0.15 to 1.2
Green loamy sand	0.6 to 0.9
Dark sandy loam with scattered pebbles and a thin white clay-layer	0.46
Green sand, slightly mottled yellow	0.6 to 0.9
Thin layer of green-coated flints	—

A section in a borrow-pit [531 253] near Elsenham, exposed during the construction of the M11, showed tuffaceous beds within the basal London Clay:

	Thickness m
GLACIAL SAND AND GRAVEL: largely removed for aggregate and fill	0.5
LONDON CLAY	
Clay, very silty, micaceous, brown and grey mottled, weathered	2 to 3
Sand, fine-grained, clayey, weakly laminated, orange, brown and greenish brown; banded appearance, with pale buff bands having bioturbated tops and bases, approximately 10 cm thick: tuffaceous horizons, four recognised in basal metre; gradational base	1.5
Sand, fine-grained, bioturbated, shelly, dark grey-brown; some glauconite grains and small rounded black flint pebbles; sharp burrowed base	1.5 to 2.0
WOOLWICH AND READING BEDS	
Clay, blue-grey and pale grey mottled at top, red and grey below; prominent shears	about 5.0

Dr R Harland of the Biostratigraphy Research Group reports that a sample from the tuffaceous beds yielded the following dinoflagellate cysts: *Alterbidinium minor* (Alberti) Lentin and Williams, *Apectodinium paniculatum* (Costa and Downie) Lentin and Williams, *A. summissum* (Harland) Lentin and Williams and *Deflandrea oebisfeldensis* Alberti, which indicate the *hyperacanthum* Zone of Costa and Downie (1976), and the *oebisfeldensis* acme of Knox and Harland (1979). A small pit [about 530 255] near the crossing of the M11 route and the railway line exposed 1.8 m of brown clayey silt resting on putty chalk. A Thanetian age was indicated by the presence of the dinoflagellate cysts, *Alisocysta margarita* (Harland) Harland and *Hystrichokolpoma mentitum* McLean.

Basal London Clay crops out in part of the Tertiary outlier west of Stansted Mountfitchet and between Birchanger and Durrel's Wood. On the south side of Durrel's Wood [5293 2478], an excavation for a bridge-pier exposed the following during construction of the motorway in 1977:

	Thickness m
GLACIAL SAND AND GRAVEL	1.8 to 2.0
LONDON CLAY	
Sand, fine-grained, clayey, laminated, with some fine-grained sandy clay and bioturbated in places; mottled orange-brown and blue-grey	0.8
Sand, fine-grained, clayey, bioturbated, mottled as above; slightly greenish and glauconitic in places; some relict lamination; small burrows (less than 2 mm in diameter); sharp base	1.8

Sand, fine-grained, medium to dark grey, with
some clayey patches; burrows less than 1 cm
across, filled with olive-brown, smooth clay;
some laminae of brown clay, reworked;
carbonaceous fragments and bivalve casts; base
not seen   about 3.0

WOOLWICH AND READING BEDS
Clay, blue, grey and red mottled   0.5

Ward (1978) has reported on fossil fish obtained from cuttings for the M11 Birchanger Interchange [514 213; 515 216] to the south of here.

FOUR

Plio-Pleistocene: Crag

During the Survey's sand and gravel resource assessments, boreholes sunk through the drift between Stebbing and Great Bardfield (Thomas, 1982) recovered marine shelly glauconitic sands, here described as 'Crag' and delineated on the published map. These sediments are probably equivalent to the Red Crag of the Suffolk coast (Harmer, 1900), which Mitchell et al. (1973) formally placed at the base of the British Pleistocene. However, much of the Red Crag of Suffolk is now thought to be of Pliocene age (Funnell, 1987). It is doubtful that Harmer's subdivisions of the Crag are applicable on a broad scale. This account follows the lines of the description of the Braintree district (Ellison and Lake, 1986), where Crag deposits were similarly discovered but not formally assigned to any particular subdivision.

The undoubted shelly Crag deposits in this district consist dominantly of pebbly quartzose sands. The sands typically have a bimodal grain-size distribution (Thomas, 1982); the gravel fraction is dominated by well-rounded flint (75 per cent), with subordinate angular flint (12 per cent) and quartz (9 per cent), some quartzite and traces of ironstone, sandstone, phosphatic nodules, igneous and metamorphic rocks. These beds are green in their unweathered state due to the presence of glauconite, but are commonly leached and oxidised at outcrop to an orange-brown colour in their upper part; tabular iron-cemented beds also occur. The maximum recorded, though incomplete thickness of Crag is 12.6 m at Great Bardfield [6972 2959].

These undoubted Crag deposits rest on a peneplain at between 60 m and 70 m above OD between Great Dunmow and Great Sampford (compare Figures 4 and 5). This peneplain rises westwards towards Thaxted, where it lies at an altitude of nearly 80 m above OD. In this area it is overlain by Crag deposits which are completely oxidised and lacking in shell detritus. Smaller outliers of pebbly sand were detected beneath the drift near Elsenham, where their base

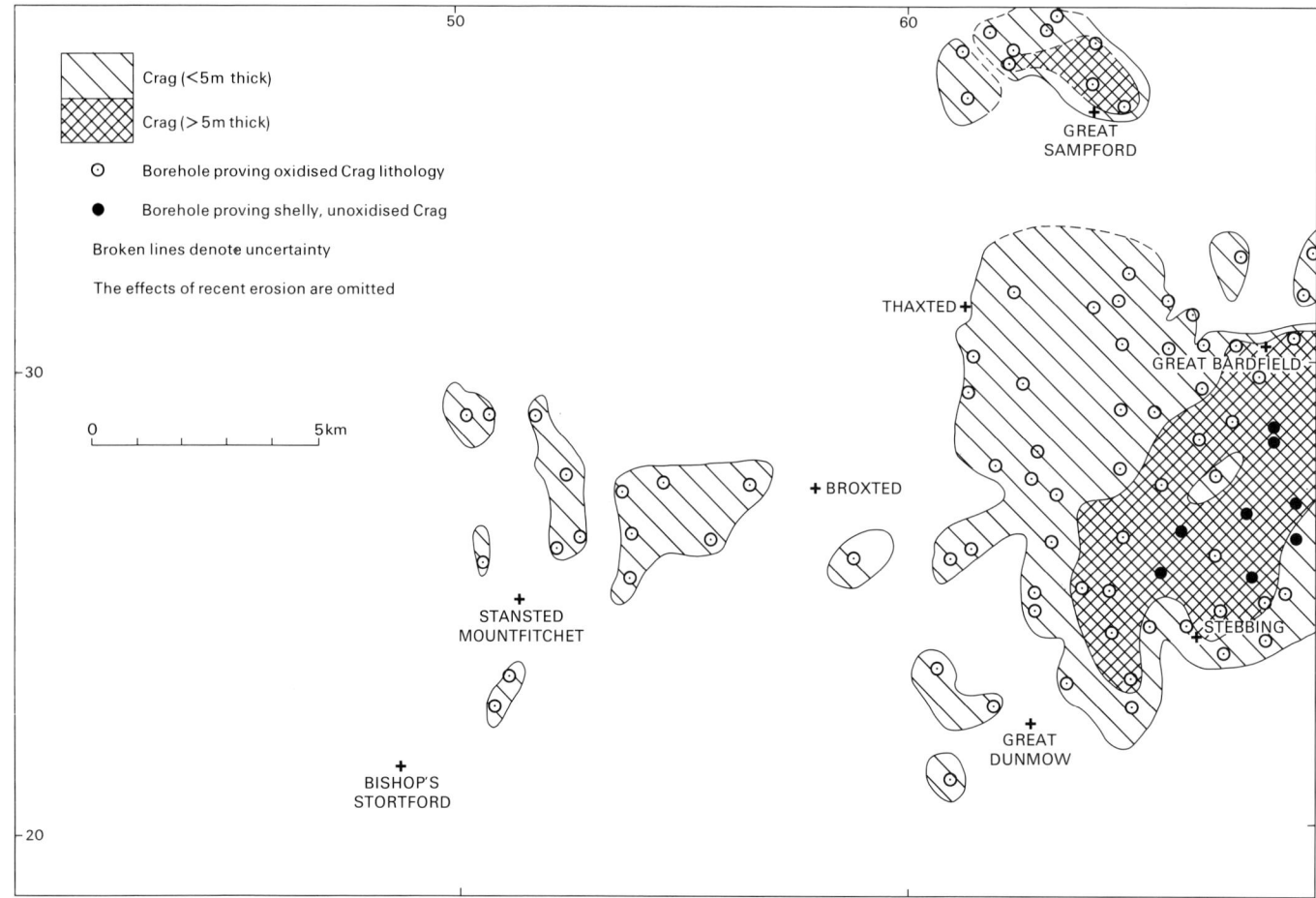

Figure 4   Generalised distribution of the Red Crag and its probable oxidised equivalents

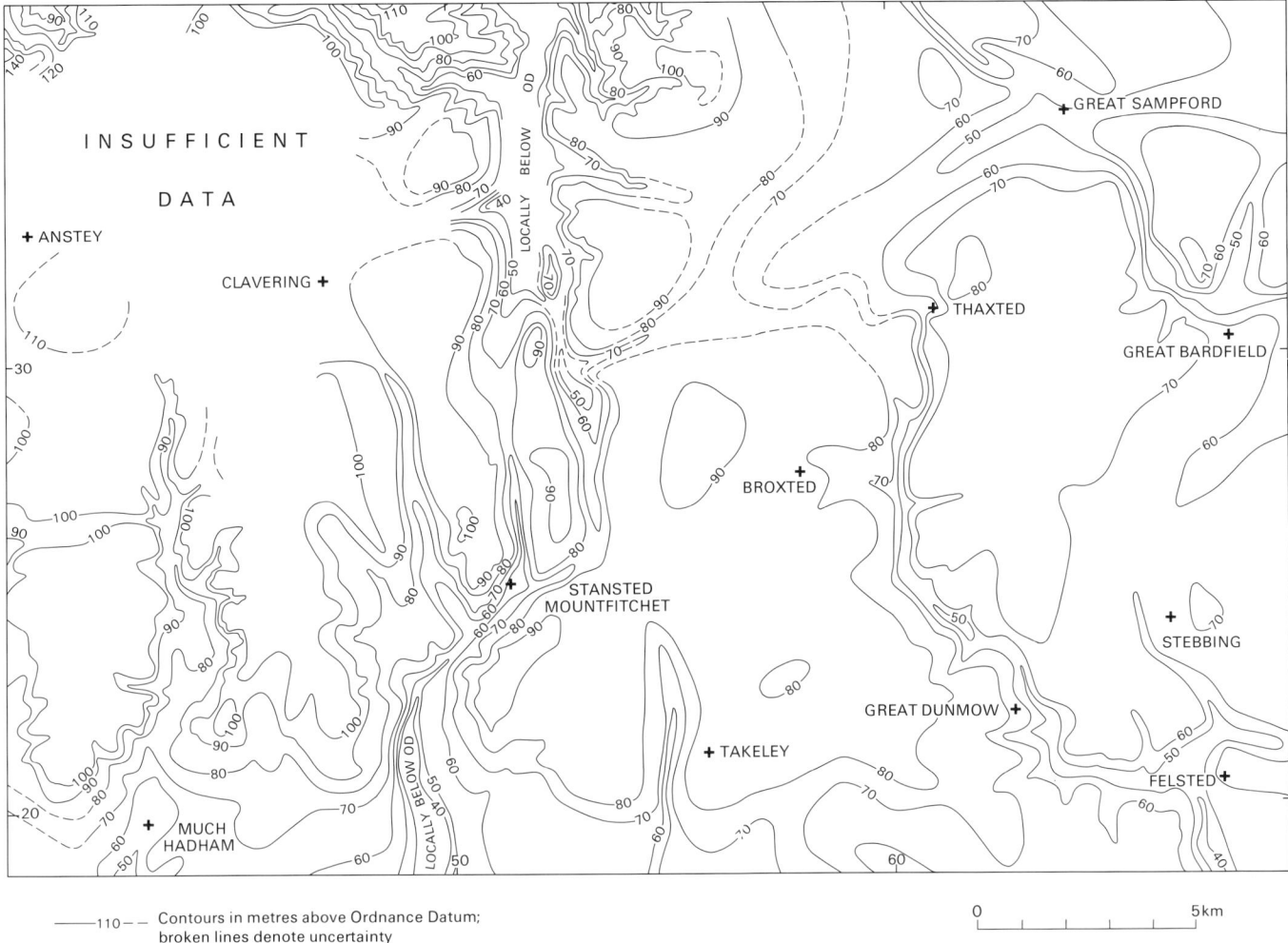

Figure 5 Contours on the surface of the Cretaceous and Tertiary strata

lies at levels up to 96.6 m above OD; similar deposits occur near Great Sampford (Hopson, 1981; Marks, 1982). The western examples are thin and may have been subsequently recycled in part, but their distribution indicates the former extent of the Crag (Figure 4). During the mapping programme, it was not possible to distinguish these lithologies from the overlying Kesgrave Sands and Gravels and consequently the Crag depicted on the published map comprises the shelly deposits only.

In a sand-pit at Elsenham [546 266] in 1985, one part of the exposed sequence (see p.22) showed ripple cross-lamination which indicates bimodal current directions and hence a tidally influenced depositional environment. The associated uniformly fine-grained sands and other comparable deposits in the district have been assigned to the Chillesford Sand Member of the Norwich Crag Formation by Mathers and Zalasiewicz (1988). In this memoir these deposits are described with the Kesgrave Sands and Gravels.

The boreholes which encountered Crag deposits are detailed in the various BGS Mineral Assessment Reports (p. iv). Boreholes at Bluegate Hall Farm, Great Bardfield [6877 2945] and Tollesburies Farm, Stebbing [6758 2677] provided microfaunas consistent with a late Pliocene/early Pleistocene age.

Thin Red Crag sediments have been identified in quarries at Hollow Lane, Widdington [530 310] by Mathers and Zalasiewicz (1988) and at Elsenham [550 265] by Hopson (1981). The first authors also described sections in workings at Cowlands Farm, Stebbing [669 233] and Canfield [578 211] which showed patches of basal pebble bed up to 0.7 m thick resting on London Clay and overlain by Kesgrave Sands and Gravels. The latter pit was dug after the completion of the geological survey.

FIVE

Quaternary: Bramertonian to Anglian

The division of the Quaternary into stages reflecting the climatic oscillations during this period is shown in Table 3. This table shows the various cold and warm climatic stages recognised in late Pliocene–Quaternary times; few of the cold intervals are thought to have reached full glacial status in lowland Britain. The sediments of Bramertonian to Anglian age form the greater proportion of the Quaternary deposits in the district. Three main units are represented: Kesgrave Sands and Gravels, Glacial Sand and Gravel and Till (Boulder Clay); Glacial Silt occurs in restricted areas. These deposits were laid down on a broadly planar surface which was probably inherited partly from that described above as marking the base of the Crag. Several buried channels which are commonly coincident with the present-day river valleys dissect this surface. The most marked are those that link the Cam and the Stort valleys (Chapter 6; Figures 11 and 12).

The two suites of gravel mentioned above, together with the Crag, constitute a single widespread sand and gravel body in north Essex. Previous field-workers tended to regard them all as 'glacial', although both Prestwich (1890) and Solomon (1935) had recognised that in Essex and Suffolk a suite of quartz-rich gravels could be distinguished from other types. More recently, workers including Hey (1965, 1980) and Rose and Allen (1977) have refined the stratigraphy of the drift deposits and established the following sequence, which is broadly applicable to the whole region:

Anglian:	Lowestoft Till	: glacial
	Barham Sands and Gravels	: glaciofluvial
	Barham Loess	: periglacial
	Barham Arctic Structure Soil	: periglacial
Cromerian:	Valley Farm Rubified *Sol Lessivé*	: humid, warm, temperate
Beestonian: (to Pre-Pastonian)	Kesgrave Sands and Gravels	: periglacial

These formation names are based on stratotypes in south-east Suffolk. Rose, Allen and Hey (1976) recognised the fluviatile quartzose Kesgrave Sands and Gravels in a broad tract through Essex and Suffolk including three localities in this district, namely Widdington [532 307], Elsenham [546 266] and Great Sampford [640 360]. Hey (1965) recognised quartzose Westland Green Gravels to the west of Bishop's Stortford, which he suggested (Hey, 1980) were equivalent to the part of the Kesgrave Sands and Gravels that occurs at higher elevations in north Essex and south Suffolk. Thus, continuity of this spread of gravel deposits was established from near Reading, through St Albans and Bishop's Stortford, and thence eastwards into Essex, marking a former course of the River Thames (Rose, 1983). The term Kesgrave Sands and Gravels has been adopted in the present account although in practice local mapping difficulties may have caused some Crag and glacial deposits to be grouped with this suite of sediments (p.15 and below). In boreholes, however, pebble counts and grading curves have generally enabled the various deposits to be distinguished.

The Lowestoft Till forms part of an extensive sheet in Essex and East Anglia. Although a two-fold subdivision was proposed by Baden-Powell (1948) it is now generally ac-

Table 3 Late Pliocene and Quaternary deposits of the Great Dunmow district

Stage		Deposits
Flandrian	w	Alluvium: floodplain deposits. Calcareous tufa Alluvium: mere deposits
Devensian	c	Head deposits. ?Most River Terrace deposits
Ipswichian	w	
Wolstonian	c	
Hoxnian	w	Lacustrine deposits
Anglian	c	Glacial deposits
Cromerian	w	
Beestonian	c	⎫
Pastonian	w	⎬ Kesgrave Sands and Gravels
Pre-Pastonian	c	⎭
Baventian	c	
Bramertonian	w	Chillesford Sand*
Antian	w	
Thurnian	c	
Ludhamian	w	
Pre-Ludhamian	c	Crag +

* the sand component of the Kesgrave Sands and Gravels as described herein; age uncertain
+ age uncertain
c: cold climate
w: warm climate

cepted that only one extensive chalky till is present in Essex and adjoining counties (see for example Perrin et al. (1973)).

It is probable that a considerable time interval separated the deposition of the Crag and the Kesgrave Sands and Gravels, but differing views of the age and depositional history of the latter have been presented. Baker and Jones (1980) suggested that their Widdington Sands (Kesgrave Sands and Gravels of this account) were early Anglian fluvioglacial deposits, whereas others (Hey, 1980; Rose et al., 1976) advocated a Beestonian periglacial origin. It is doubtful whether an early Anglian age can be sustained in the light of field evidence, though some deposits may represent sands and gravels recycled during the Anglian glaciation. A Beestonian age is linked to the recognition of a widespread rubified clay-rich horizon at the top of the Kesgrave Sands and Gravels. Rose et al. (1976) postulated this to be a *sol lessivé*, formed during the warm temperate Cromerian Stage (although see subsequent discussion).

KESGRAVE SANDS AND GRAVELS

The Kesgrave Sands and Gravels are a composite facies array occurring at differing elevations with respect to the sub-Crag peneplain. The deposits are sheet-like in form and crop out on the valley slopes of some of the major rivers and many of their tributaries. Their distribution largely follows that of the Tertiary outcrop and has been determined from sand and gravel assessment boreholes in the areas of National Grid sheets TL 41, 51, 52, 61, 62, 63; elsewhere they are thin or absent (Figure 6). West of Bishop's Stortford (in the area of 1:10 000 sheets TL 42 SW and SE), contours on the subdrift surface are approximately parallel to the valley sides and here it is difficult to distinguish between in-situ deposits and their soliflucted derivatives; as a result, the outcrop pattern may give a false impression of the thicknesses of the deposits. The base of the Kesgrave Sands and Gravels is typically sharply unconformable on the Tertiary bedrock, but where Crag intervenes, the contact may be difficult to define because of the effects of recycling.

Two broad facies equivalents of the Kesgrave Sands and Gravels can be recognised within the district. In the south, well-sorted, yellowish brown to grey, pebbly, predominantly medium-grained, quartzose sands and sandy gravels occur (Figure 6) which are lithologically similar to the Kesgrave Sands and Gravels of the type area. The gravels contain flint, quartz, quartzite and some sandstone pebbles and are up to 10 m thick. The deposits to the west of Bishop's Stortford (the Westland Green Gravels of Hey (1965)) are of a similar composition but are thinner. Few large exposures

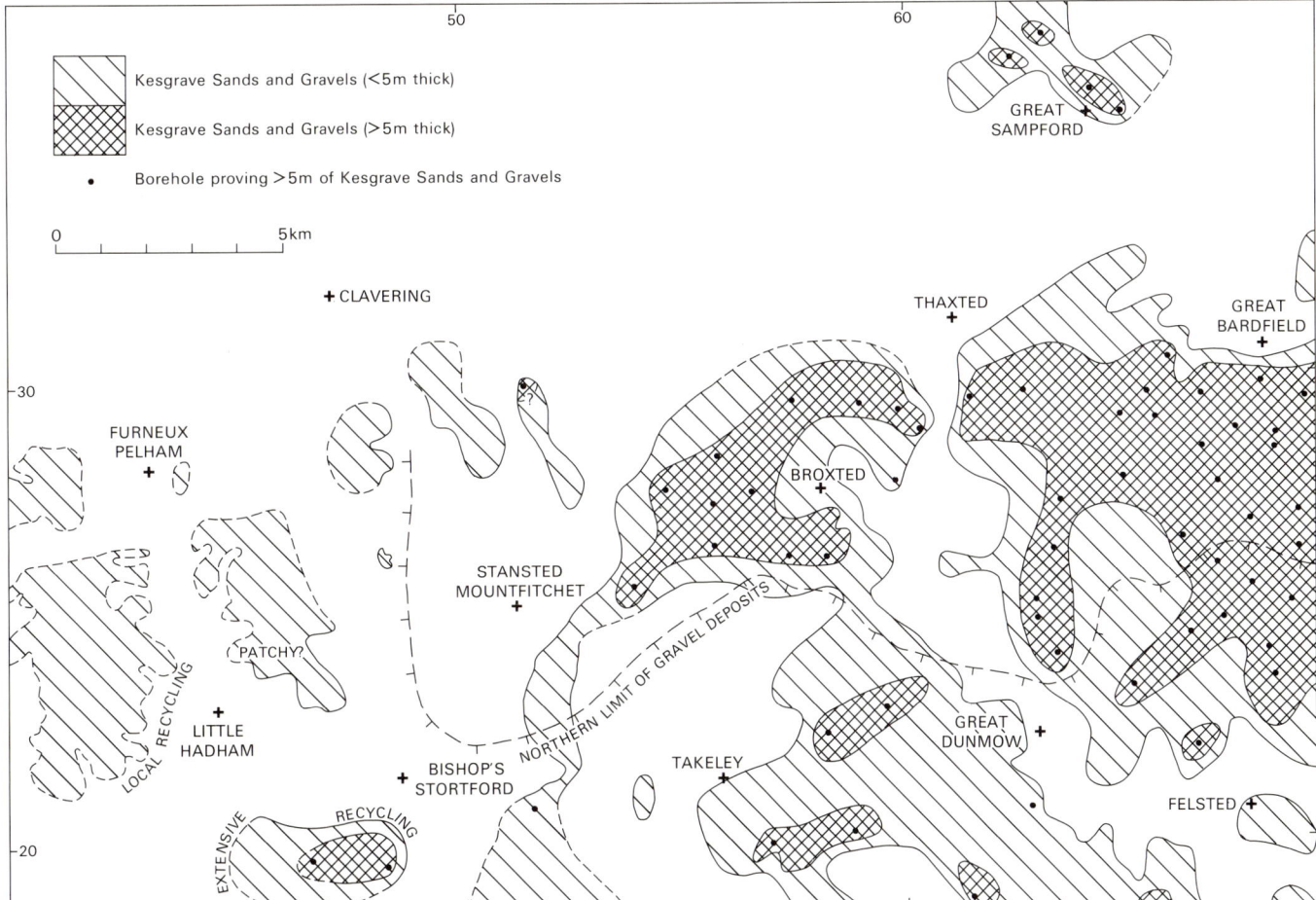

Figure 6 Generalised distribution of the Kesgrave Sands and Gravels

have, however, been recorded in this area; Hey (1965, p.408) described the deposits at the type locality [422 215] as 'poorly bedded, poorly sorted...' and was clearly concerned as to whether the sediments in general were in situ; it is considered that large parts of these deposits have been disturbed by geliturbation and solifluction.

In the northern part of the district, the deposits mapped as Kesgrave Sands and Gravels are very well-sorted, fine- to medium-grained sands, up to 15 thick, with a markedly unimodal particle size distribution (Marks, 1982). Thin seams of sandy gravel occur locally; the pebbles are predominantly of flint, with some quartz and ironstone. The distribution of these sands is similar to that of the Crag (Figure 4); at Elsenham [546 266], bimodal current directions were obtained from ripple and dune size cross-stratification in these deposits (Mathers and Zalasiewicz, 1988). These authors attributed the structures to a tidally influenced environment and assigned these, and comparable sediments, to the Chillesford Sand Member of the Norwich Crag Formation (of possible Bramertonian age).

In the central part of the district, these two facies of the Kesgrave Sands and Gravels overlie the westward extension of the major peneplanation surface to the east. In the vicinity of Bishop's Stortford a low, south-facing bluff delimits the southern extent of the peneplain (Figure 5); gravels are commonly absent on the bluff-slope which grades down to a more gently inclined south-facing bevel. Two episodes of deposition are suggested by the presence of these subdrift bevels, but the intervening bluff cannot be traced far eastwards.

Involuted red, brown and grey mottled clayey sands and pebbly sandy clays, typically up to 1.5 m thick, occur at the top of Kesgrave Sands and Gravels sequences in pits at Widdington [532 307], Great Sampford [640 360] and Stebbing [669 233] (Frontispiece), and in boreholes between Great Easton [607 254] and Felsted [676 203]. They form a rubified *'sol lessivé'* which has been interpreted (Rose and Allen, 1977) as the lower part of a fossil soil profile, the upper part of which suffered illuviation and the subsequent overprint of disturbance by ground-ice (the Barham Arctic Structure Soil of Rose and Allen). At Great Sampford (p.25) the formation of the red clays demonstrably predates the effects of glaciotectonics; at Stebbing (p.24; Wilson, 1983) the variegated clays appear to line the walls of gulls (voids formed by periglacial cambering processes). Rose and Allen (1977) suggested that the soil was formed in a humid, warm temperate climate although they accepted Catt's reservations (1977) about possible diagenetic changes in buried soils and the validity of their climatic interpretation. Wilson (1983) pointed out that the structures at Stebbing may have provided potential sites for the preservation of the upper (eluviated) part of the soil horizon but its absence throws doubt on the genetic processes involved.

GLACIAL SAND AND GRAVEL

These glacial outwash deposits crop out principally in the Cam and Stort valleys, but boreholes have shown that they are also extensive in the eastern part of the district (Figure 8). They generally comprise brown and grey, medium- to coarse-grained, poorly sorted, clayey sands and gravels containing angular and nodular flints together with clasts of chalk, sandstone, limestone and weathered igneous rock; chalky sands and finely laminated silt lenses are common. These deposits differ from the Kesgrave Sands and Gravels in their poorer sorting, greater lithological variation and the presence of large amounts of locally derived and nondurable material. Up to 14 m are present near Ugley Green [52 29] (see also Chapter 6, p.38).

The unweathered deposits commonly show planar bedding, channel-fill structures, large-scale cross-bedding and intraformational ice wedges. Within the upper 2 m or so of the weathered zone, the effects of decalcification have destroyed the bedding, and collapse structures are locally present. An occurrence on the south side of the valley of Debden Water was found to contain interbedded chalky solifluction breccias close to the southern margin [5358 3399] of a presumed channel-fill.

In the Ugley Green area, the sands and gravels are clearly related to a large, well-defined channel beneath the till (Figure 7 and p.38). Near the margin of the channel, structureless sands contain involuted clay masses. Elsewhere in the district, and particularly in the east, gravels occur both within (up to 2 m thick) and at the base of the till. The latter examples are commonly related to channels in the bedrock topography. Figure 8 shows this association; the distribution plot is based largely on information from sand and gravel assessment boreholes which have provided a more refined stratigraphy than that depicted on the published geological map.

TILL (BOULDER CLAY)

Till is the most widespread drift deposit in this district. It occurs over much of the higher plateau area and, to a lesser extent, in glacial channels at lower elevations. The typical lithology is a pale brown and grey, very chalky, sandy clay with clasts of chalk, limestone, ferruginous sandstone, subangular and nodular flints, and subordinate vein-quartz, quartzite and igneous and metamorphic rocks. The till is generally firm to stiff, overconsolidated, structureless and is probably largely a lodgement till laid down beneath a major ice sheet during the Anglian glacial period. In the upland areas it is typically up to 20 m thick.

Perrin et al. (1973) showed that in Essex and Hertfordshire the till contains 38 per cent of calcium carbonate in the matrix (<2 mm) and, regionally, that chalk clasts comprise 56 to 84 per cent of the whole by weight. Perrin et al. (1979) also produced trend surface analyses of the various grain sizes and of the heavy minerals present. In the weathered zone, the till is decalcified and leached to a dark brown stony clay to depths of 1.5 m or less.

The base of the till is generally sharp and broadly undulating. There is commonly a basal unit, up to 1 m thick, of locally brown, chalk-free and crudely laminated clay. Beds of laminated silt and glacial sand and gravel occur within the till, particularly in the north-eastern part of the district (Figure 8). Locally, poorly consolidated tills have been encountered near the margins of the buried channels; the origin of these is uncertain.

Figure 7 Ugley sand quarry: section in north-east corner

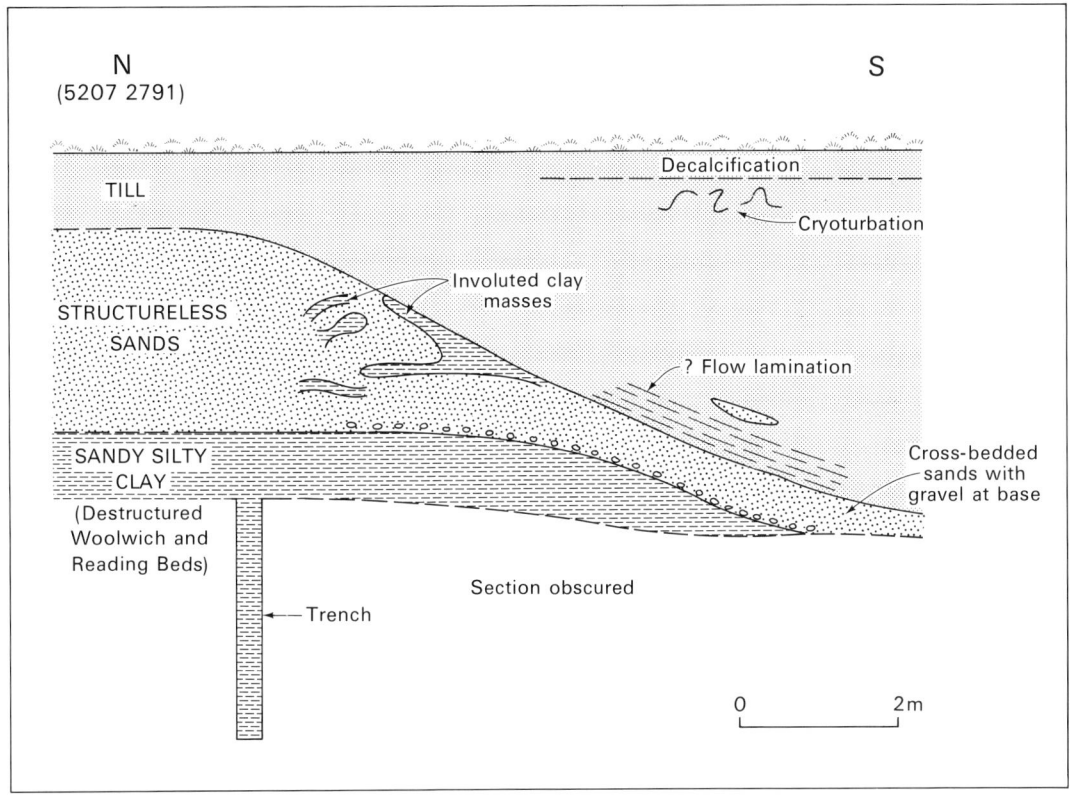

GLACIAL SILT

These beds comprise buff or bluish grey calcareous silts and silty clays with lenses of breccia, rare dropstones and chalky sand and gravel intercalations. Nodules of 'race' occur within the weathered zone. The silts are commonly finely laminated with some small-scale cross-bedding, but localised crumpling and brecciation of the laminae or complete de-structuring, resulting from ice action, occurs in places. Glacial Silt occurs in the Cam and Stort valleys (Chapter 6), and laminated beds have been proved in boreholes as thin units within the boulder clay elsewhere (Figure 8). In the Cam and Stort valleys, their thickness varies considerably in response to the bedrock topography; in the Cam valley 65 m were proved within 90 m of drift in a borehole [5163 3645] near Wendens Ambo. On a small scale it is probable that both the variable thicknesses and the lateral discontinuity of these silts reflect relationships between ?stagnating ice and the surrounding sediments.

The glacial silts reflect subaqueous slackwater conditions and the presence of rare dropstones indicates the localised presence of floating ice. The recognition of destructured silts near Ugley Green and elsewhere suggests that some of the Glacial Silt was deposited prior to the final ice-advance which disrupted it, perhaps in an ice-dammed lake. Conversely, the extensive presence elsewhere of undisturbed beds indicates that these were laid down later in the waning phase of the glaciation or, less likely, that permafrost enhanced their resistance to subsequent glacial disruption.

INTER-RELATIONSHIPS OF THE DEPOSITS

The most typical sequence of deposits in the district is (in ascending order): Crag, Kesgrave Sands and Gravels, Till; the last locally contains lenses of gravel and of silt. In places the rubified *sol lessivé* and glaciofluvial sands and gravels (Rose et al., 1976) occur between the till and the sands and gravels, but they are not generally mappable, and are recognised only in exposures.

There is a close relationship between the form of the bedrock surface (Figure 5), which reflects the subglacial topography, and the overlying stratigraphy. The bedrock surface is generally planar beneath interfluves, where the standard succession noted above is characteristic, but is channelled beneath certain valleys, the channels being infilled by varied glacial sequences (see Chapter 6).

Locally, only lenticular masses of Kesgrave Sands and Gravels are preserved beneath the till; on valley slopes these gravels pass downslope into their soliflucted derivatives (termed Head Gravel). Outcrops of Tertiary deposits occur not only in the floors of the main tributary valleys but also as 'windows' in the heads of minor valleys and on topographic saddles.

DETAILS

The River Ash valley

Kesgrave Sands and Gravels were exposed in a small, abandoned, 5.5 m-deep pit north of Gravesend [4416 2672]:

CHAPTER FIVE QUATERNARY: BRAMERTONIAN TO ANGLIAN

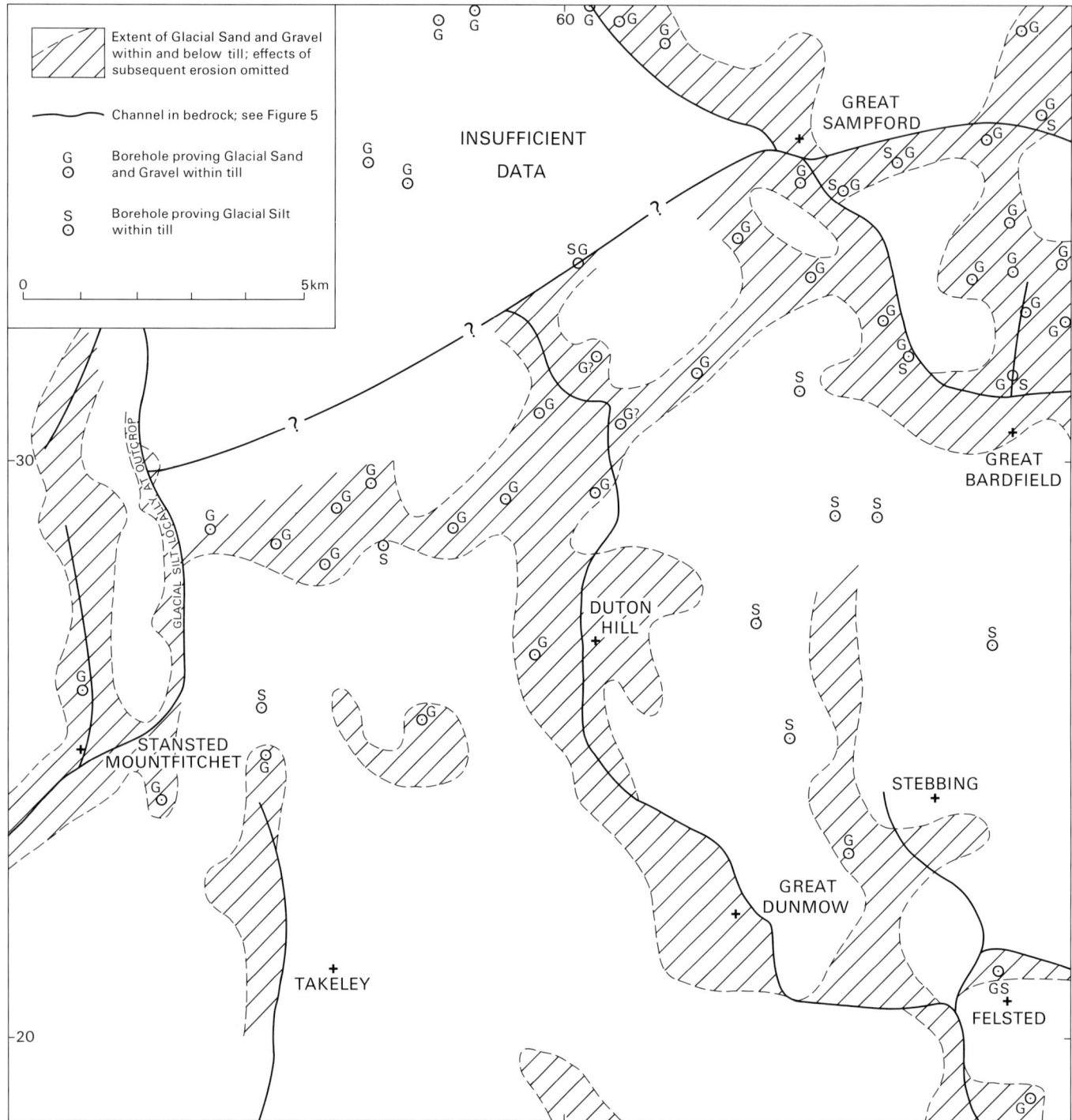

Figure 8 Generalised distribution of the Glacial Sand and Gravel and Glacial Silt in the eastern part of the district

	Thickness m
Top soil	0.3
HEAD GRAVEL	
Gravel, very clayey, unsorted, orange, with a wide size range of angular and rounded pebbles mainly of flint. Draped on bed below	1.0
KESGRAVE SANDS AND GRAVELS	
Gravel, poorly sorted, orange to brown, matrix supported with sandy clay; most pebbles less than 5 cm in diameter, well-rounded and dominantly flint, with some quartz and quartzite (?soliflucted)	2.0 to 3.0
Sand, medium-grained, buff to yellow, cross-bedded, with 'sharp' sand grains and clay galls	0 to 1.0

?WOOLWICH AND READING BEDS
Silt, slightly clayey, grey, with no trace of
lamination — about 1.0

The Kesgrave Sands and Gravels were also exposed in a small working east of Albury Hall [4331 2532]:

	Thickness m
Top soil	0.45
KESGRAVE SANDS AND GRAVELS	
Gravel, sandy, dark grey and brown, predominantly of flint clasts but with some well-rounded quartzite pebbles; matrix supported by buff, fine-grained sand; cross-bedded, with beds of buff, medium-grained sand	1.50

Available borehole logs add no significant lithological details to those of the Kesgrave Sands and Gravels described from exposures. The maximum thickness of these deposits recorded from boreholes is 8.1 m at a site [4136 2116] near Bromley.

The following section was measured in a deep ditch south of Chaldean's Farm [4230 2032], below 0.5 m of soil:

	Thickness m
HEAD	
Clay, silty and sandy, red to orange, with abundant mainly small pebbles	1.52
Gravel, orange to red, with sandy clay matrix. Clasts angular to subangular, predominantly flint	0.50
GLACIAL SAND AND GRAVEL	
Sand, medium-grained, orange with red patches; slightly clayey towards the top	0.56

Large, possibly lensoid masses of soft, putty-like chalk have been recognised in the till outcrop south of Hadham Ford and south-west of Albury Hall. These occurrences may represent irregular masses of very chalky till or rafts of chalk carried as erratics by the ice sheet.

In the Farnham Green area [473 253] complex relationships between the drift and solid formations suggest that diapiric structures may be present. Irregular interfaces may be more extensive in this area than the surface geology suggests.

The River Cam valley

This section details the occurrences of Kesgrave Sands and Gravels, both in situ and reworked; other deposits associated with the buried channel are described in Chapter 6.

A trial pit [5272 3352] to the west of Ringers Farm encountered material which probably represents collapsed Kesgrave Sands and Gravels and other materials in a solution hollow in Chalk. Here, 0.8 m of till rested with a planar, horizontal base on 1.6 m of orange-brown and buff, generally structureless, clayey, coarse-grained sand with numerous large flints and indistinct wispy laminae locally. In a nearby pit [5271 3354] till rested directly on Upper Chalk.

Two trial pits were sunk to the east-north-east of Quendon church to determine the nature of the sands which occur on the higher ground. The first [5190 3073] proved 1.9 m of medium- to coarse-grained sand beneath 0.5 m of top soil. The sands were buff, grey and reddish brown, and crude ?bedding was defined by clayey sand partings and local developments of clay-ironstone pellets. No other structure was apparent and the beds were probably reworked.

The second pit [5194 3071] showed the sequence:

	Thickness m
Top soil: sandy clay, dark brown with flinty wash	0.6
KESGRAVE SANDS AND GRAVELS	
Sand, medium- to coarse-grained, clayey, structureless, dark brown, with scattered pockets of brown, slightly sandy clay and yellow sand; scattered angular flints; abrupt gradational base	0.8
Sand, fine- to medium-grained, orange, yellow and grey, micaceous, well-bedded or laminated; bedding disturbed by minor faults (including ?thrusts), contortions and injection structures; local lenticles of brownish grey, micaceous, clayey silt; generally in situ	1.8
RELICT TERTIARY DEPOSITS	
Silt, clayey, glauconitic; intercalations of fine-grained sand and scattered, black-coated flints	0.1
UPPER CHALK	
Chalk, white, hard	—

The thicker unit of sand is regarded as Kesgrave Sands and Gravels, and the unit above as a reworked derivative, comparable with the sands encountered in the previous trial pit.

Three faces in the pit to the south-west of the church were re-excavated. One face [5165 3058] showed the following:

	Thickness m
Fill and chalky till; irregular base	about 2.0
KESGRAVE SANDS AND GRAVELS	
Sand, medium-grained, moderately well-sorted, yellow, buff and orange-brown variegated, with irregular lenticles of rounded medium flint gravel; indistinct bedding (?colour banding); grey and red mottled, slightly sandy, clay lenticles with ironstone nodules	0.55
Gravel, medium, with rounded to subangular flints and subordinate vein quartz; orange and yellow, poorly sorted, coarse sand matrix	0.15

A nearby face [5167 3063] showed a comparable sequence of:

	Thickness m
Fill and chalky till; irregular base	0 to 0.5
KESGRAVE SANDS AND GRAVELS	
Sand, fine- to medium-grained, micaceous, yellow to orange-brown	0.4 to 0.9
Gravel, medium-grained, sandy, moderately sorted; coarse sand matrix; dominantly rounded flints with ?ferruginous coating, and subordinate quartz pebbles (less than 5 per cent); finer-grained towards base	1.1

In the south-western corner of the pit, a face [5158 3058] showed:

	Thickness m
Top soil, fill and Head	0.7
TILL	
Clay, sandy	0.7
KESGRAVE SANDS AND GRAVELS	
Sand and clay, red mottled, reworked	0.45

	Thickness m
Sand, reddened in part, with clay lenticles	0.25
Gravel, medium, rounded, fairly well-sorted	—

In each of these faces, the lowest unit is thought to be in-situ Kesgrave Sands and Gravels.

The following section in the Kesgrave Sands and Gravels was exposed in the south-east corner [5318 3066] of the Hollow Lane Gravel Pit, Widdington:

	Thickness m
HEAD	1.0
TILL, chalky	1.0
KESGRAVE SANDS AND GRAVELS	
Sands, medium-grained, well-sorted, reddish brown and white, cross-bedded, with scattered thin gravel stringers	8.0
WOOLWICH AND READING BEDS	0.6
UPPER CHALK	—

More recent exposures enabled Mathers and Zalasiewicz (1988) to subdivide the sands in this pit as follows: Kesgrave Sands and Gravels, 2 m, resting on Chillesford Sand, 1.5 m, on Red Crag, 1.5 m.

Farther south, near North Hall, a trial pit [5214 3005] revealed the following sequence:

	Thickness m
Top soil	0.3
TILL	
Chalky boulder clay, silty, sandy in parts, olive-grey, with scattered angular flints; abundant chalk fragments	0.7 to 0.9
Clay, silty to finely sandy, olive-green with red and brown mottling; scattered rounded flints; transitional base; passing down to	0.15 to 0.35
KESGRAVE SANDS AND GRAVELS	
Sand, coarse-grained, pebbly, structureless, buff, orange and reddish brown mottled, with slightly clayey finer patches; pebbles predominantly of flint and vein quartz (pebble count: 196 flint, 46 quartz, less than 10 other types); indistinct, fairly irregular base	about 1.15
TERTIARY DEPOSITS	
Clay, silty, olive-green with reddish brown mottling; poorly defined ?bedding picked out by mottling, but otherwise structureless; infrequent small rounded flints; becoming more uniform downwards; passing into green, ?glauconitic, clayey, fine-grained sand; irregular base	about 0.35
Flint horizon: large, nodular, glauconite-coated flints	0.05
UPPER CHALK	
Chalk, decalcified to dark reddish brown clay in upper 0.03 m; white, hard chalk below	0.3

Much of this material probably infills a solution hollow in the Chalk. On the east side of the trench, till was seen to be pocketed into an irregular Chalk surface at about 1.5 m depth. The Chalk surface was inclined at 45°W. The pebbly sands (reworked Kesgrave Sands and Gravels) and underlying mottled clays (of Tertiary origin) were probably slumped into the solution hollow.

Elsenham and Birchanger area

On the east side of the valley at Elsenham, the stratigraphical relationships were investigated in a trial pit [5356 2758] at Elsenham Nursery which showed:

	Thickness m
Wash: sandy, flinty clay	1.0
TILL	
Sands, medium- to coarse-grained, and sandy clay with scattered rounded flint pebbles, mottled orange, buff and brown, structureless	0.7
KESGRAVE SANDS AND GRAVELS	
Sands, fine- to medium-grained, clayey in west face; some quartzite pebbles; finer-grained at base; cross-bedded clean sands in east face	0.6

An extension of this pit westwards demonstrated that the sandy basal till passed laterally into a chalky and more clayey lithology. The latter had a sharp, slightly irregular base that was seen to be inclined gently valleywards.

The Elsenham Sand Pit [546 268] exposed up to 6.5 m of ochreous and pale-grey, cross-bedded, fine- to coarse-grained sands, overlain by up to 6.5 m of chalky till (Plate 3). Pale grey clay galls with reddish rims were present in the lower beds. Small-scale channel structures were common. The upper sand beds were pebbly (quartzitic) and contained clay partings and lenticles of brecciated clay. The contact with the till was seen to be planar but this surface was locally modified by involution structures (load-casting) and fluid extrusion. In places the basal metre of the till was distinctly flow-banded and contained locally derived Tertiary clay material. The sands are assigned to the Kesgrave Sands and Gravels on the published map but, as noted previously (p.15), more recent observations suggest that the lower part of the sequence may be Crag. Mathers and Zalasiewicz (1988) recognised at least 3.5 m of Chillesford Sand, of uniform fine-medium grain size, beneath 3 m of pebbly Kesgrave Sands and Gravels and, at the base, following Hopson (1981), about 1 m of iron-stained, poorly sorted, locally pebbly Red Crag.

The railway cutting [536 263] at Elsenham appears to be largely excavated in Glacial Sand and Gravel. In contrast, a trial pit [5381 2623] to the east, near Elsenham Cross, encountered Kesgrave Sands and Gravels, which were probably slightly reworked:

	Thickness m
HEAD	
Loam, dark brown, silty, with abundant angular flints	0.7
KESGRAVE SANDS AND GRAVELS	
Sands, fine-grained, well-sorted, weakly planar-bedded, buff, iron-stained; local subvertical ?liesegange staining; no pebbles seen	1.3
Sands, fine- to medium-grained; seeping water	0.7

Two pits were dug in the Parsonage Farms area, south of Stansted, to investigate the lithologies of the local gravels. The more westerly [5162 2389] showed:

	Thickness m
Top soil: flinty loam, dark greyish brown	0.5
GLACIAL SAND AND GRAVEL (reworked)	
Gravel, medium to coarse, clayey, sandy, unsorted, structureless, noncalcareous, ?soliflucted; mainly	

Plate 3 Elsenham Sand Pit. Pale grey, very chalky till overlies Kesgrave Sands and Gravels (A 13548)

	Thickness m
nodular, angular and rounded flint cobbles up to 0.2 m diameter; scattered pockets of coarse sand; poorly defined, ?cryoturbated, gradational base	1.6

LONDON CLAY (?reworked basal beds)
Clay, silty to fine-grained sandy, brown and pale grey mottled with a greenish tinge, soft, micaceous in parts, with glauconitic patches; structureless to ?weakly laminated; included ?cryoturbated flints from above in upper part   0.8

The second pit [5170 2365] to the south-south-east and at a higher elevation exposed:

	Thickness m
Top soil: dark sandy loam	0.5

GLACIAL SAND AND GRAVEL
Sand, medium-grained, clean, fairly well-sorted, variegated grey, white, orange and reddish brown, slightly clayey in parts, with scattered rounded flints; gradational base   0.55
Sand, coarse-grained, dark reddish brown to orange with pale grey patches, structureless, with rounded flint pebbles and sporadic quartzites common throughout; pebbles more common towards fairly sharp base   0.60
Sand, fine- to medium-grained, slightly clayey, buff and reddish brown, ochreous; scattered small rounded flints; fairly sharp irregular base   0.10
Sand, coarse-grained, clayey, poorly sorted, gravelly, with rounded flint pebbles throughout and large shattered flints locally; more clayey downwards; irregular base   0.25

LONDON CLAY
Clay, slightly silty, pale bluish grey with orange-brown mottles, micaceous, structureless; injected into unit above in east end of face   0.3

This pit was subsequently extended eastwards to show:

	Thickness m
Clay, sandy, olive-grey mottled, slightly flinty; (?degraded till)	1.0
Sands (?flow-banded)	0.6
LONDON CLAY	—

Both pits encountered Glacial Sand and Gravel but, in the latter, recycled Kesgrave Sands and Gravels were evidently present.
 In the south-west corner [5183 2176] of a borrow-pit at Start Hill, the sequence was:

	Thickness m
TILL	about 1.0

KESGRAVE SANDS AND GRAVELS
Gravel, predominantly rounded but some subrounded; flint, quartzite and vein quartz pebbles up to 13 cm diameter but most less than 10 cm; vaguely bedded; medium- to coarse-grained sand matrix   1.0
Sand, medium- to coarse-grained, white to buff, with stringers of rounded flints, vein quartz and quartzites   0.5

Workers at the pit reported a maximum thickness of 2.5 m for the gravel deposits, and London Clay was scraped from the base of the pit.

The Kesgrave Sands and Gravels were also exposed in a disused brickpit at Tilekiln Green [522 213]. At the western end [5213 2130], 1.4 m of bedded sands and sandy gravels were exposed in 1979. The gravel content was similar to that in the borrow-pit but rare mottled green volcanic pebbles and one rotted granodiorite were noted; black flints of Tertiary derivation were also present. Low angle cross-bedding was visible and some of the sand units were graded. In the southern face [5223 2128], 100 m from the previously described exposure, gravels dominated the 2.2 m exposed; the gravel beds were internally structureless, and quartzite and vein quartz pebbles comprised 25 per cent of them; thin sand lenses less than 4 cm thick were present.

The River Chelmer valley

On the north side of the disused railway line, 1 km north-north-east of Armigers, a small gravel pit [5992 2995] exposed the following:

	Thickness m
TILL	
Clay, silty, pale fawn and brown, with abundant chalk fragments; calcrete at the base	0.7 to 2.0
GLACIAL SAND AND GRAVEL	
Gravel of flint cobbles, crudely horizontally bedded, with a sandy matrix; some brown sandstone and vein quartz pebbles	1.4
Sand, chalky, well-bedded, with some layers of fine to medium chalk gravel	1.8

A channel-fill of gravels, which cut down through to the base of the section, displayed tightly synclinal bedding indicating collapse subsequent to deposition. Within the channel the sequence was:

	Thickness m
GLACIAL SAND AND GRAVEL	
Gravel (forming core of structure), comprising angular and nodular flint cobbles, and brown sandstone and white vein quartz pebbles in an ochreous brown, chalky, silty sand matrix; internally structureless	up to 2.1
Gravel, medium, chalky, but decalcified in places, crudely bedded; ochreous brown to dark brown clayey sand matrix	0.8
Sand, well-bedded, with much chalk; thin lenses of chalky gravel	0.9
Sand, medium-grained and fine gravel, cross-bedded, pale ochreous brown to dark brown; chalky but decalcified in part	1.1

South of this section a further pit, partly back-filled, showed the following section in glacial deposits in the upper part of its face [5970 2989] at the western end of the pit:

	Thickness m
TILL	
Clay, sandy, structureless; scattered small, angular flints with long axes horizontal in basal part; passing down into	0.4
Clay, gravelly; matrix as above; subhorizontal, subrounded to nodular flints up to 15 cm in size; sharp irregular ?cryoturbated base	0.4 to 0.7
GLACIAL SAND AND GRAVEL	
Gravel, clayey, chalky, pale ochreous to creamy brown, with angular to subrounded flints up to 5 cm in diameter	0 to 0.2
Sand, medium- to coarse-grained, slightly clayey	0 to 0.2
TILL	
Clay, silty, slightly sandy in places, structureless, with scattered flints and chalk pebbles; trace of bedding given by parallelism of long axes of pebbles. (?Flow till)	0.6

A small gravel pit [5965 2895] 200 m east of Armigers showed:

	Thickness m
Gravelly, loamy soil	0.3
GLACIAL SAND AND GRAVEL	
Gravel, sandy, clayey, dark brown, with angular flints; trace of bedding in places but generally cryoturbated; thin beds of sand up to 4 cm thick; pipes of this material extended into material below; sharp base with iron pan	1.8 to 2.9
Gravels with subordinate sands; gravels chalky, with rounded chalk fragments up to 4 cm; also angular and nodular flints up to 10 cm; coarse bedding indicated particularly by sandy layers; medium- to coarse-grained chalky quartz sand with fine chalk gravel layers up to 14 cm thick; sands laminated and cross-laminated; gravel decalcified adjacent to pipes from bed above	2.9

Two phases of activity are represented in this section; firstly, deposition of the glacial chalky gravels and sands at the base and, secondly, either the remobilisation of those beds or deposition of new material by solifluction. The gravel in the upper deposit was markedly more angular than in the unit below. In addition to chalk and flint, clasts in the gravels included quartzites, sandstones, shelly Jurassic limestones, clay-ironstone nodules and pebbles of rotted basic igneous rock and metamorphic quartzite.

The Stebbing Brook valley

In the gravel pits [669 233] at Cowlands Farm, Stebbing, the following section, 2.8 m high, was recorded in a disused face (A in Figure 9) [669 234] which showed a number of laterally persistent units in the Kesgrave Sands and Gravels, locally disturbed by large gull-like superficial structures (enlarged tension fissures induced by mass-movement processes):

	Thickness m
Top soil: pale brown pebbly loam	0 to 0.3
KESGRAVE SANDS AND GRAVELS	
Silt and silty clay, grey with orange and red mottling (rubification); becomes slightly coarser downwards; sporadic gravel lenses (up to 3.0 m × 0.5 m), in deep reddish brown sandy clay matrix towards base of unit, contained abundant subrounded to rounded patinated flints (up to 0.06 m diameter) with fairly common, smaller, well-rounded white quartz and very rare dark (?igneous) pebbles; load structures probably due to cryoturbation are present beneath the gravel lenses and at the transitional base of the unit	0 to 1.0
Sands, medium-grained, well-sorted, pale greyish brown with orange-brown mottling,	

finely laminated in parts; sporadic cross-bedded lenses defined by thin bands of fine gravel; scattered larger rounded flints; becoming coarser downwards and passing into gravels towards the base; the gravels comprise subangular and subrounded patinated flints, small rounded white quartzes, very sporadic cherts and igneous rocks (highly weathered), and may form up to half the thickness of this unit; fairly distinct but irregular base — 0.4 to 0.8

Gravel, fairly coarse, with predominantly rounded to subangular flints, and subsidiary rounded white quartzes, quartzites and scattered weathered ?igneous pebbles in an orange-brown coarse sandy matrix — 0.05 to 0.3

Sands, fine- to medium-grained with some coarser bands, orange-brown, red and grey; fine lamination and cross-bedded laminae present, disturbed by small-scale normal faults; scattered gravel lenses and bands containing small flakes of brown silty mudstone (10 to 20 mm long) (?London Clay); gravels more common in western end of face — up to 1.0

Gravel, as above, alternating with thin beds of brown, medium- to coarse-grained sand — up to 1.2

The beds within this section were seen to be disturbed by two large V-shaped structures at either end of the face. Their dimensions on the face (not necessarily true dimensions) were about 7 m wide at top (5 m wide at base) and 2.8 m deep. Bedding thinned and wrapped around the structures, which were filled with a mélange of gravel in a reddish brown clayey sandy matrix. Remnants of the silty clay seen at the top of the pit face occurred within the structures and retained laminations defined by red and grey colour banding. There was some alignment of pebbles at the margins of the structures.

Another exposure (Face B in Figure 9) showed the sequence:

Thickness
m

KESGRAVE SANDS AND GRAVELS
Sands, grey and red mottled (rubified) with a basal gravel parting — 1.2
Sands, white and pale grey, fine- to medium-grained — 2.5
Gravel, sandy (similar to Face A) — up to 3.0

At the northern end of Face B, the sands and gravels were seen to be channelled into. About 20 m of the channel floor was exposed and the following section of the channel-fill was recorded:

Thickness
m

TILL
Clay, silty, mottled pale grey and orange-brown — up to 1.0
Clay, silty, with scattered flints, mottled dark greyish blue and brown (decalcified boulder clay) — up to 0.3

KESGRAVE SANDS AND GRAVELS
Sand, fine-grained, clayey, with orange and red mottling; irregular lenses of gravel — up to 3.0

This exposure and the nature of the periglacial structures were discussed in Rose (1983, pp.149–162).

The Pant valley

The former sand pit at Great Sampford was temporarily cleared by BGS in 1980. The section in the northern face [6410 3608] is shown in Figure 10. Because the beds were highly variable, a number of sections were examined (A to F in Figure 10). Section A showed the following succession:

Thickness
m

GLACIAL SAND AND GRAVEL
Sand, coarse-grained, clayey, slightly chalky, pale grey and brown mottled, with irregular clay skeins — up to 0.3
Sand, fine-grained, silty, pale grey with slight brown mottling; increasingly silty downwards; sharp base — up to 0.5
Sand, medium- to fine-grained, orange-brown, with a few minute flint chips; irregular ?faults picked out by pale streaks; trace of weak bedding; ?increasing clay bind towards base; ferrocrete, 0.1 m thick, at base — up to 1.06

?RUBIFIED SOIL HORIZON
Clay, very silty, with scattered medium sand grains; ochreous and pale grey mottling in upper part (up to 0.34 m thick); distinct red zone below (0.14 m thick); pale grey distinctly sandy zone at base (up to 0.1 m thick); sharp base — up to 1.61

KESGRAVE SANDS AND GRAVELS
Sand, fine- to medium-grained, buff to pale grey, distinctly laminated in upper 0.1 m; olive to brown clayey silt lenticles below — 2.91

Section B was as follows:

Thickness
m

GLACIAL SAND AND GRAVEL
Gravel, stained orange-red and grey, medium-grained, matrix supported; pebbles up to 3 cm diameter, mostly rounded, but a few angular to subangular; dominantly of flint, with minor vein quartz and white quartzite; matrix of coarse-grained, poorly sorted clayey sand: gradational base — up to 0.1
Sand, coarse-grained, sparsely pebbly, conspicuously striped in grey, orange and brick-red (stripes oblique to bedding), poorly sorted, with rounded to angular grains; pebbles of flint, vein quartz and quartzite, mostly rounded: gradational base — up to 0.4
Gravel, medium-grained, sandy, mottled and striped in grey, orange-red and brick-red, apparently matrix-supported; pebbles mostly rounded, up to 5 cm (commonly 1 cm diameter), of flint, vein quartz and white quartzite, with minor ?chalk; matrix of slightly clayey coarse sand: base irregular and sharp — up to 0.7
Sand, medium-grained, moderately well-sorted, orange and grey irregularly (vertically) striped, with some red mottling near top; a few lenses, 1 to 3 cm thick, of coarse-grained sand — 2.9

Figure 9a Plan of gravel pit at Cowlands Farm, Stebbing

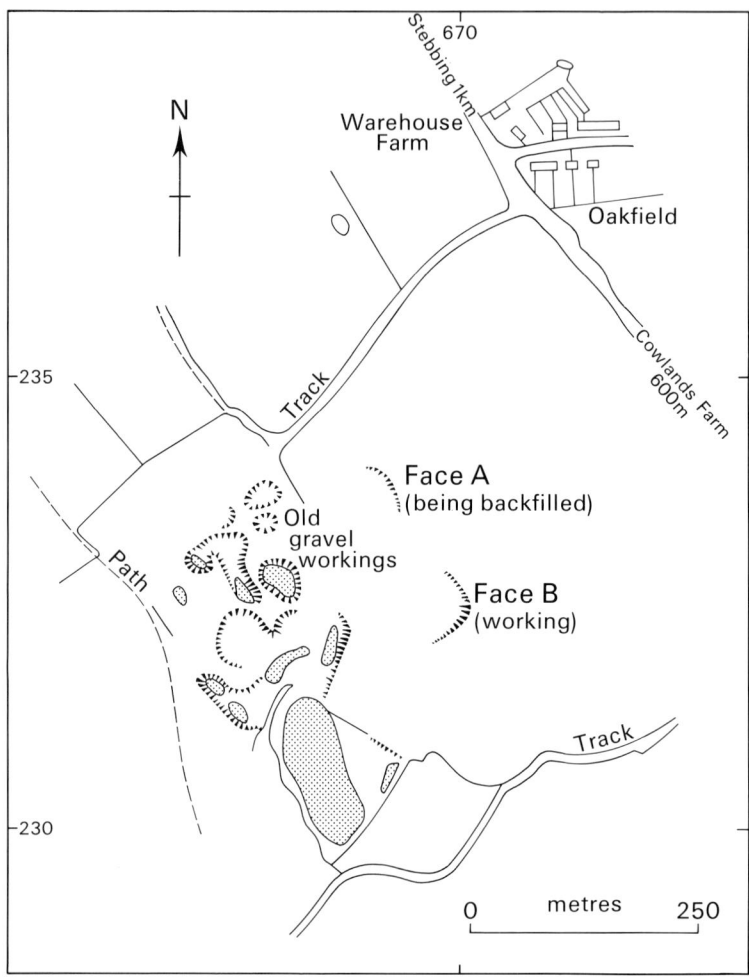

Figure 9b Periglacial structures in face B of the pit

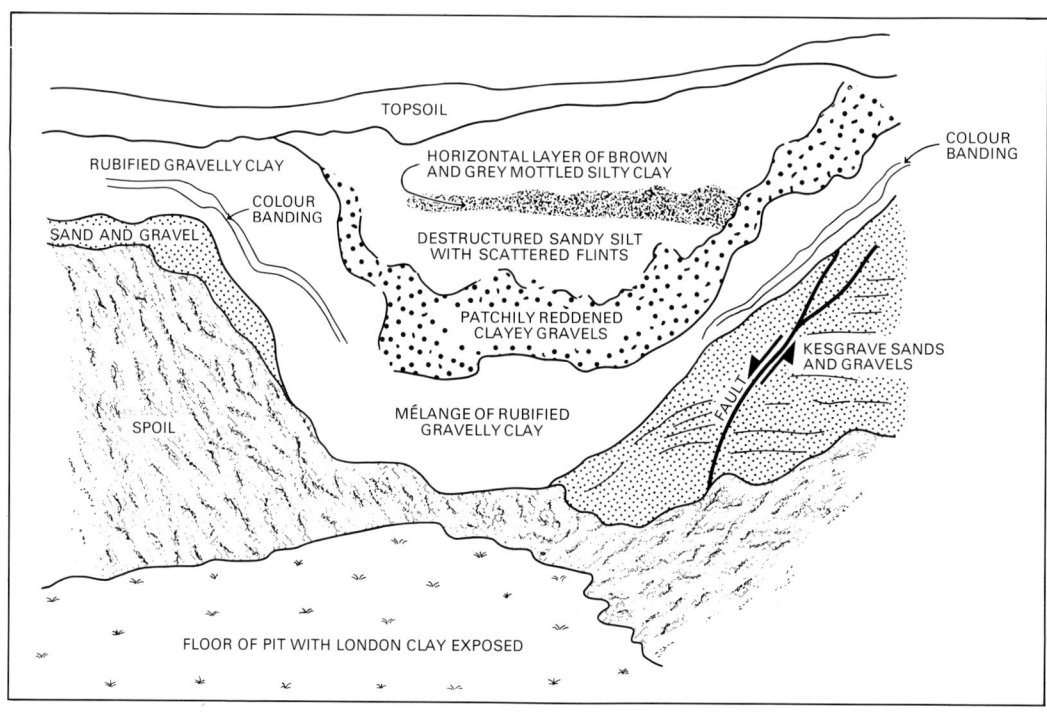

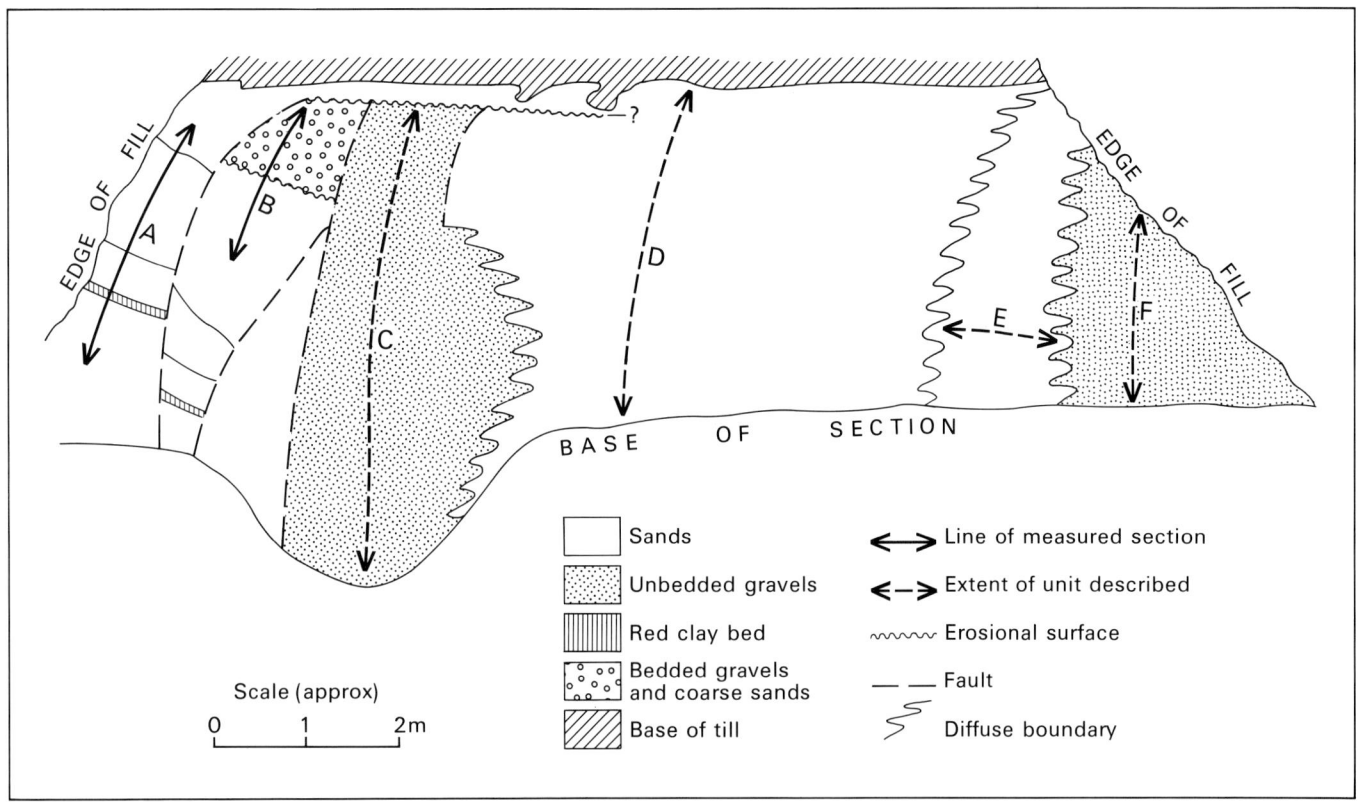

Figure 10 Sketch section of north face of sand pit at Great Sampford

In Section C, orange, medium- to coarse-grained, sandy gravels contained rounded to subrounded pebbles of flint, vein quartz and white quartzite with minor ?acid volcanic rocks. Section D comprised buff, almost structureless, silty, fine-grained sands. The sands in Section E were similar in texture but contained red clayey zones and showed a faulted and highly deformed internal structure. Unbedded clayey gravels with silty intercalations were present in Section F. The lithologies in sections C to F are apparently restructured Kesgrave Sands and Gravels, in the main. A further section [6406 3609], some 20 m west of the above face, showed up to 4 m of undisturbed, cross-bedded, pebbly, medium-grained sands (Kesgrave Sands and Gravels) beneath till.

The disturbances seen here are probably glaciotectonic in origin. The alignment of the reverse faults at the western end of the figured exposure suggests that pressure was exerted from the north, the direction from which the Anglian ice-sheet advanced. The merging of the gravels with the sands on the eastern side of the main gravel wedge, together with the massive structure and vague hints of strong contortion in those sands, suggest that the advancing glacier locally induced mixing and flow in the underlying sediment. The gravel in the main wedge was possibly solifluctued and reworked.

SIX

Quaternary: Anglian glacial channels

INTRODUCTION

Deep glacial channels, termed buried tunnel-valleys by Woodland (1970), were first recognised as such by Boswell (1914); earlier writers had described anomalously thick drift deposits beneath some of the valley floors of East Anglia, but favoured a subaerial mode of erosion. Woodland compared the channels with the tunnel-valleys of Denmark and the Rinnentaler of north Germany; he reviewed the well records that proved the East Anglian examples and suggested that earlier valleys had been modified by subglacial processes comparable with those responsible for the European examples. Since 1970 few writers have referred to these channels (see Cox, 1985), although they are considered to represent the major drainage routes for the Anglian meltwaters, which may account for the general paucity of outwash materials beneath the till sheet.

There appear to be two distinct sizes of glacial channels in southern East Anglia, namely the deep channels, catalogued by Woodland (1970), and much shallower features, which are probably present in many of the other modern valleys. The deep glacial channels identified in the region can be classified as radial or tangential from their position in relation to the former ice margin and the till sheet.

a) RADIAL ICE-MARGINAL CHANNELS

Channels of this type are aligned approximately normal to the supposed ice-front, near its margin; they formed part of a radial subglacial drainage network. They include one of the better documented examples of a deep buried channel, namely that of the Gipping valley in Suffolk, which is narrow, 'probably bifurcating with clear evidence of four or five scoured basins separated by thresholds' (Woodland, 1970, p.530), with drift extending down to a level of more than 30 m below OD and with rockhead at more than 45 m below OD in three water supply wells.

b) TANGENTIAL ICE-MARGINAL CHANNELS

Several examples of this type of channel are recorded between Chelmsford and Colchester; they lie subparallel to the Danbury–Tiptree ridge, which appears to have effectively impounded the ice sheet, and include the channel occupied by the Marks Tey interglacial deposits (Turner, 1970; Ellison and Lake, 1986). The deposits within these channels tend to be dominantly fine grained, and they incorporate tills and laminated sediments.

c) TANGENTIAL, INTRA-TILL SHEET CHANNELS

Channels such as that of the Stour to the north-west of Sudbury fall into this category because they lie within the main till sheet and are aligned approximately at right angles to the inferred direction of ice movement. These examples appear to be largely filled with silts.

d) RADIAL, INTRA-TILL SHEET CHANNELS

Although this is the most common type according to Woodland (1970), the Cam–Stort buried channel is the only major one in Essex known to occur within the area of the till-sheet and to be aligned approximately parallel with the ice-flow direction. Other smaller-scale channels occur, however. The Cam–Stort system appears to be unusual because it crosses the Chalk escarpment. The sediments within this channel system are very variable (see below).

In Essex and East Anglia the deep channels generally lie beneath the alluvium of the valley floors and have influenced the modern drainage pattern. Recent erosion has commonly removed any evidence for the form of the bedrock surface on the flanks of the depressions. In the Cam–Stort system, however, where dissection has been less intense, cross-sections show that rockhead has an undulating profile which steps down towards the central deep 'core' of the channel system.

THE CAM–STORT BURIED CHANNEL

Morphology and fill of the channel complex

Varied drift sequences occur in the valleys of the rivers Cam and Stort; the distribution of the deposits beneath the main till sheet is shown in Figure 11. This till and later deposits have been omitted to show the inter-relationships of the earlier drifts more clearly. Baker (1976) has described low-level diamictons, which he thought were derived from the plateau till by earthflow and debris sliding in the later Devensian stage. However, the size and configuration of these deposits suggest that they lie on an irregular, step-like bedrock profile on the flanks of the drift complex (Figure 12a); in part, at least, they probably predate the plateau till. Woodland (1970) concluded that the buried channel of the River Stort, which contains more than 60 m of drift deposits, 'is probably little more than 400 m wide and is clearly steep-sided'. Later borehole information and various cross-valley profiles have confirmed that, within the Cam–Stort system, the valleys are locally steep-sided. Figure 12 utilises field and borehole data from adjacent areas and, though partly interpretative, illustrates the steep-sided valley profile and the relationships of the various drift deposits. However, boreholes are too scattered to prove the presence of 'thresholds' in the geomorphological sense, or to delimit the precise subdrift valley profile.

The important features of the Cam–Stort buried channel are two deep drift-filled complexes, each of which is continued southward by an outwash channel, filled with Glacial Sand and Gravel, incised into bedrock through a thin and localised capping of Kesgrave Sands and Gravels. The detailed morphology and fill of these drift complexes is largely

Figure 11 Morphological and lithological elements of the Cam–Stort buried channel complex beneath 'Plateau till'

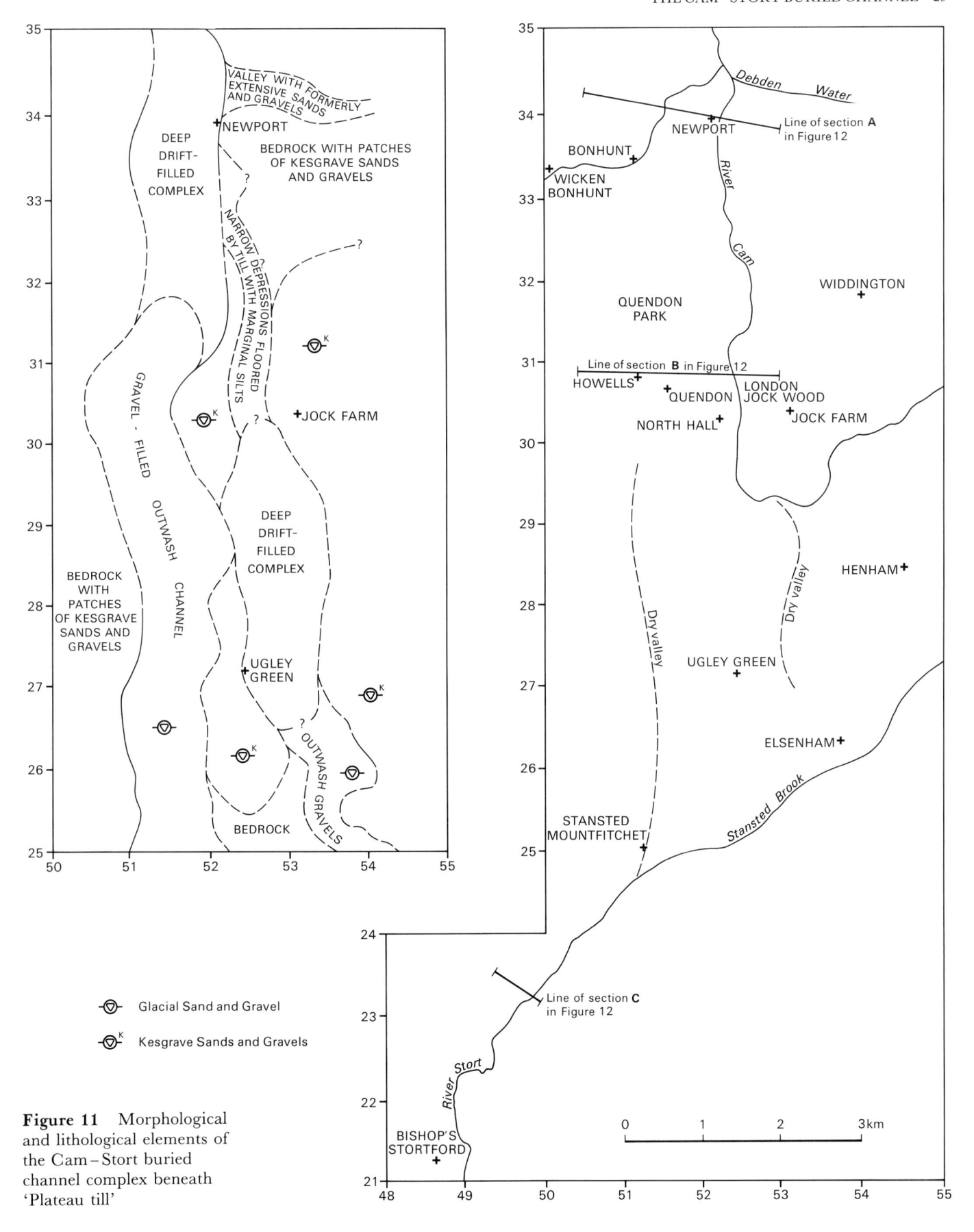

unknown. Boreholes between Newport and Elsenham have proved variable thicknesses of drift deposits in the channels; 103.6 m were proved in one borehole [5199 3439] of which 78.9 m were 'sandy loams' (?Glacial Silt). Boreholes at Newport [5217 3365] and Quendon [5212 3139] proved mainly till whereas others encountered sequences of clays, silts and gravels, or silts and sands, or sands and clays. In the Bishop's Stortford area the maximum depth proved to bedrock is 61.9 m [4901 2105]. There appears to be no systematic variation of lithology within the channels, but the occurrences of Glacial Silt serve to indicate the approximate extent of these features.

Deposits of Glacial Silt crop out extensively to the south-east of Bonhunt Water, between Bonhunt [5115 3345] and Newport, and occur on the flanks of the River Cam valley between Newport and London Jock Wood [5260 3055]; there are patches farther north towards Littlebury, and they are also present on the western side of the dry valley between North Hall and Ugley Green. At Bonhunt trial pits [around 5113 3332] have shown that disrupted laminated clays underlie till; nearby motorway boreholes have proved thick, though apparently laterally restricted sequences (up to 20 m) of silts and clays beneath a thin cover of till, for example at [5162 3316] and [5316 3342]. The laminated clays invariably rest on, and are locally interbedded with sand and gravel deposits which were not completely penetrated. Trial pits and boreholes have proved similar sequences within the buried channel as far south as Quendon Park [5222 3139], and sections of destructured silts and clays, overlain by till, were observed in a sand quarry at Ugley [5200 2795].

Undisturbed well-laminated silts are sporadically exposed on the slopes of the present-day River Cam valley between [5238 3215] and London Jock Wood [5262 3045]. The configuration of the solid and drift deposits in this area suggests that the silts were deposited in an elongate depression which was flanked generally by Chalk bedrock but locally by till (Figure 12b). Immediately north-west of the Widdington gravel pit, the silts abut (at 64 m above OD) against the Kesgrave Sands and Gravels and Chalk [5277 3120] with a steep-sided contact. Near London Jock Wood a trial pit [5236 3075] proved the silts resting on Upper Chalk at 75 m above OD; close to the valley floor [5250 3087] the silts rest on an irregular Chalk bedrock surface at 68 m above OD.

To the west of the River Cam, trial pits and boreholes have shown that the silts are absent and that till blankets the Upper Chalk down to and below alluvium level. Farther north, however, the silts occur on the western side of the present-day River Cam [524 330]; here their steep-sided junction with Chalk bedrock defines the western margin of the depression. Tills, resting on Upper Chalk, occupy the lower ground to the east of the River Cam, repeating the relationships to the south. Till deposits between North Hall [5225 3030] and Jock Farm [5310 3040] are thought to be dissected remnants of a drift feature which may have restricted the southward extension of the lake in which the silts were deposited. The maximum elevation of 84 m above OD for the silts near the Widdington gravel pit coincides with the regional planation level of the bedrock surface in this area and may reflect the near-maximum level of siltation within the channel.

At several localities, apparently well-bedded, heterogeneous sequences occur, particularly in the upper parts and towards the flanks of the channels. The sediments generally comprise either interbedded tills, solifluction flows and gravels, or interbedded silts and sands. These sediments interdigitate in a complex fashion, with abrupt lateral changes of lithology, which may be related to ice-contact phenomena. For example, a trial pit [5197 3252] near Newport exposed 1.6 m of undisturbed laminated clays with subordinate sands overlying 0.5 m of coarse gravels, yet till was proved only a few metres to the west.

All grades of sands and gravels are present; the deposits range from poorly sorted, coarse, clayey and sandy gravels to well-sorted, fine-grained sands. The thicker deposits tend to occur either at the bottoms or towards the flanks of the channels. Extensive glacial sands and gravels with interbedded silts and clays underlie till in the buried channel between Newport and Quendon, as proved in motorway trial boreholes. The proportion of fine-grained sediments in these deposits is unusually high; they may be subglacial in origin or may represent outwash deposits which have been over-ridden, and presumably reworked, by the advancing ice. Gravels are present in the Quendon area, extending southwards to Bishop's Stortford; they have a markedly erosional base and contain lenses of till. In Quendon Park [516 316] they are also interbedded with till. In a trial pit [5114 3086] at Howells, they appear to overlie brown flinty clays with basal flint lag gravels, which probably represent localised solifluction deposits at the ice margins.

South of Quendon the gravels fill a discrete channel, possibly an overspill, which effectively links the River Cam buried valley system with that of the River Stort. The northern limit of the gravel-fill is apparently quite abrupt, whereas to the south the gravels merge imperceptibly with those of the River Stort system. Near the flanks of this channel, reworked Kesgrave Sands and Gravels have been recognised within the sedimentary sequence. These sediments display weak bedding structures, only moderate sorting and clayey lenses. These features were noted in temporary exposures at Elsenham [5381 2623] and Quendon [5165 3058; 5194 3071]; at the last-noted locality the sands exhibited dewatering structures and minor faults.

Glacial sands and gravels are present in the valley of Debden Water to the east of Newport, where they form a narrow discontinuous outcrop confined to the lower valley slopes close to alluvium level. The gravels have occasionally been worked from small pits down to, and locally below alluvium level, and they probably extend beneath the present-day alluvium and Head. A trial pit within old gravel workings on the south side of the valley [5358 3399] proved over 2.5 m gravels, solifluction gravelly clays and chalk breccias, the base of which was not penetrated. An excavation in the southern face of the workings revealed a wall of shattered chalk, either in situ or little moved, which defines the southern limit of the deposit. The gravels may be the exhumed relics of a formerly extensive deposit which filled the valley and was probably connected to the main buried channel complex at Newport.

The cross-section of the drift deposits in the River Stort valley to the north of Bishop's Stortford is shown in Figure

Figure 12 Cross-sections of the Cam–Stort buried channel complex

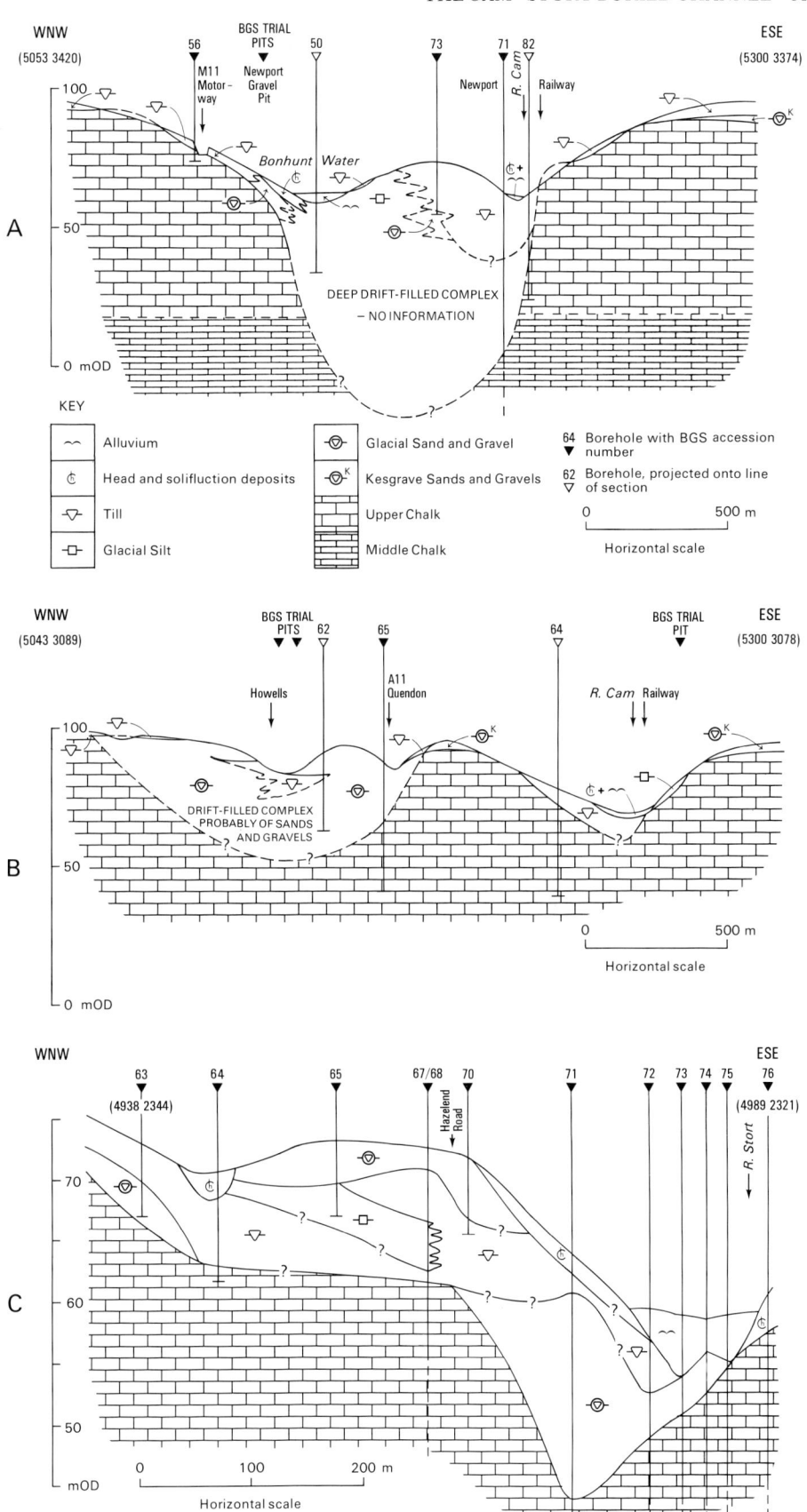

12c. To the south of this profile, the sedimentary fill of the buried channel largely comprises silts and clays, locally underlain by gravels. The section probably lies near the southern limit of the gravels that extend southwards from Quendon. The nearby patch of glacial silts near Foxdells [488 228] possibly has a steeply inclined contact with the till to the north.

Two boreholes at a site near the Railway Hotel, Bishop's Stortford [c.4917 2106], 50 m apart, proved 14.3 m and over 26.1 m of drift deposits. These and other boreholes at this site variously proved complete gravel sequences (of which the uppermost are terrace deposits) or terrace gravels overlying organic shelly clays with 'lignite', which in turn rest on gravels. A summary of boreholes proving buried channel deposits near Bishop's Stortford is given in Table 4.

Sedimentary relationships within the Cam – Stort buried channel

Many of borehole logs that have demonstrated anomalously thick sequences of drift deposits within the Cam – Stort buried channel lack adequate lithological descriptions. This is largely because of the difficulties of drilling through water-saturated sediments and of the sampling techniques employed. In his regional appraisal of well-documented boreholes, Woodland (1970) quoted a number of logs which showed complex sequences including alternating clays and sands; he concluded that 'the fill seems invariably to consist of bedded sequences of sands, gravels, silts, loams, brick-earths and clays'. This conclusion led to the belief that most of the sediments were deposited subaqueously and that 'in the lower reaches of the present valleys where the tunnel-valleys lie well below the level of the regional boulder clay spread' ... 'the clays with chalk' ... 'are likely to resemble' ... 'laminated clays' ... 'and are not therefore boulder clay'.

More recent information from trial pits and boreholes in the Cam and Stort valleys indicates that the relationships between the sediments are even more complex: they show rapid lateral variations, including interdigitations or sharp subvertical contacts between till and other lithologies. Near the margins of the channels some sequences are complicated by the presence of ill-sorted sediments representing interbedded solifluction flows.

The environment of deposition of the sediments described above is by no means fully understood. It is likely that both lodgement tills and flow tills are represented in some of the boulder clays recorded (see also Baker, 1977). The presence of laminated silts and chalky clays may reflect subaqueous deposition, but it is not clear to what extent these sediments were laid down subglacially. Where laminated silts comprise the bulk of the channel fill, a proglacial or late-phase melt-out environment of deposition is suggested. Field relationships near Ugley (p.38) indicate that destructured silts may have been deposited proglacially and subsequently over-

Table 4 Boreholes proving the buried channel of the River Stort (arranged in order north to south)

Borehole number TL442SE/	Grid Reference	m OD	Depth to bedrock (Upper Chalk)	Nature of fill
110	4940 2301	c.64	28.3	Silts overlying interbedded gravels and ?tills (beneath Head)
94	4875 2157	60.0	c.15.2	—
3	4895 2129	—	12.0 +	Sand and gravel (beneath alluvium)
4	4895 2125	—	13.1 +	Mainly sand and gravel (beneath alluvium)
15	4888 2128	56.1	30.2	Mainly sand and gravel (beneath alluvium)
95	4901 2105	57.6	61.9	Sand and gravel
97	4905 2109	56.1	25.1	Mainly sand and gravel
98/118	4916 2104 (series)	c.61	14.3 – 26.1 +	Mainly silts and clays beneath surface gravel
99	4914 2092	58.1	16.9	Mainly sand and gravel
100	4901 2085	58.5	41.1	—
1	4907 2086	—	6.1 +	Sandy clay and gravel (beneath alluvium)
2	4908 2084	—	6.1 +	Sandy clay and gravel (beneath alluvium)
101	4889 2079	57.9	27.7	Sand and gravel (beneath alluvium)
103	4896 2069	57.9	43.9	Sands, gravels and clays
104	4909 2062	57.9	51.8	Sands, gravels and clays
5	4919 2050	—	6.1 +	Sand and gravel (beneath alluvium)
6	4928 2025	—	9.1 +	Mainly sand and gravel (beneath alluvium)
106	4935 2026	57.3	12.2	—

ridden by the ice. The varied occurrences of the sands and gravels suggest that these were deposited in a number of environments, such as gravel trains down-valley from the ice (Woodland, 1970), the fill of spillways, as subglacial channel-fills or late-phase supraglacial stream sediments, especially in the later melt-out stage. A schematic reconstruction (Figure 13) of the stages in the development of the Cam–Stort buried channel shows the relative position of these sand and gravel bodies and their relation to the other drift deposits.

The recognition of complex drift sequences within the Cam–Stort and other valleys has important implications for the glacial stratigraphy and chronostratigraphy of the region. Many of the discussions on glacial stratigraphy in recent years have focussed on the supposed recognition of more than one till in East Anglia. This supposition has, in turn, led authors to erect complex chronologies for the drift sequence. Although the bipartite nature of the Lowestoft and Gipping tills has now been generally discredited (Bristow and Cox, 1973; Perrin et al., 1979), the monoglacial approach to the remaining stratigraphical problems has been modified by the recognition of more than one till in localised areas. Thus Clayton (1957) recognised two tills in both the Chelmsford and Harlow areas. In the former the stratigraphy was shown to be in error (Bristow and Cox, 1973), but in the latter, various authors (Clayton and Brown, 1958; Baker and Jones, 1980) have suggested that there is evidence for two ice-advances, which caused successive modifications to the drainage pattern. Baker and Jones (1980) listed sites where there was evidence for more than one till (i.e. a till below the main Lowestoft till sheet). In the Harlow area, however, and at a number of the other sites, a buried channel is known or suspected to be present nearby and the complex sedimentary sequences present may, therefore, relate directly to the interbedding of glacial deposits within or on the flanks of the channel and not to any successive advances and retreats of the ice.

In his review of the 'tunnel-valleys' of Norfolk, Cox (1985) recognised several phases of channelling which he ascribed to non-glaciogenic fluvial activity. Quartzose gravels in one borehole at Dunston Common, which he compared with the Kesgrave Sands and Gravels, and therefore regarded as pre-Anglian fluvial deposits, were thought to provide evidence of a fluvial rather than subglacial origin. This isolated example may, however, be a result of reworking by glacial meltwater.

Evolution of the Cam – Stort buried channel

It is significant that in East Anglia all the major buried channels, including that of the Cam–Stort, lie on the Chalk outcrop or are believed to reach the Upper Chalk locally, even where it lies beneath relatively thick Tertiary strata. Woodland (1970, p.556) noted this and attributed the alignment of the channels to the fact that the Chalk, being 'heavily fissured, was more responsive to downward cutting than to lateral erosion'. He also noted one example in east Jutland of the 'sudden deepening of the floor of the tunnel valley' where 'the Gudenaa valley passes for a distance of about 3 kilometres from Tertiary strata on to Chalk'. He considered 'it is significant, therefore, that where tunnel-valleys in Denmark pass on to Chalk terrain, they appear to have narrow, deep, downward extensions, wholly filled with glacial sediments'.

The erosion of the channels in East Anglia has been attributed 'mainly to the gravel load carried by the subglacial streams rather than to the water itself' (Woodland, 1970, p.523), although the latter was 'under great hydrostatic pressure' (Charlesworth, 1957, p.66). Similar examples of buried channels in North Germany (Ehlers and Linke, *in press*), which are between 200 and 300 m deep and 2 to 4 km wide, and which are thought to be exclusively of Elsterian age, are filled with meltwater deposits (silts and sands); tills have only been recorded in a few places, although their presence is not always discernible. Individually, the channels show aggradation of coarse-grained deposits at their proximal ends to finer-grained sediments at their distal ends; rafts of (Tertiary) bedrock also occur. The authors ascribed the excavation of the channels to episodic and catastrophic meltwater activity at the ice-margin. Thus the development of a hydrostatic head is probably the most important cause of the erosion of deep buried channels, but it is further suggested that the contribution of phreatic water in the Chalk, particularly in the area of the Tertiary crop, was probably also significant. At the Chalk outcrop, the groundwater may have been confined below a permafrost zone, leading to increased flow of groundwater in fissures. The erosional process may have been enhanced by this means when the excavation of the channel had become well advanced.

The channels may therefore define zones where increased hydrostatic pressure was present beneath the ice sheet. Although zones of enhanced pressure may initially have been randomly distributed, with the resultant subglacial erosion preferentially developed at these sites, it is thought more likely that the required hydrostatic head would be created only in localities where the ice was considerably thicker than elsewhere. Increased thickness could be caused in a number of ways:

1 By filling of a pre-existing valley which might be either a preglacial valley (Woodland, 1970) or one excavated by proglacial meltwater, for example a spillway.

2 By overthrusting of ice caused by a down-glacier obstacle to ice movement, such as a range of hills or the combination of ridge and valley, with the latter on the up-glacier side.

The radial, intra-till sheet channel (type d), of which the Cam–Stort system is the only large-scale example, probably results from two connected processes. First, the ice filled the pre-existing Cam valley and then, probably after the main ice-sheet had advanced to the contemporary watershed, a spillway drained southwards into the Stort system. After the excavation and deepening of the latter by meltwater, the advancing ice was probably funnelled into this channel and further overdeepening resulted.

DETAILS

The River Cam valley

A trial pit [5206 3860] at Nursery Lodge, Littlebury, showed that the Head deposits largely mask the occurrence of Glacial Silt hereabouts:

Figure 13 Stages in the development of a buried channel. I early Anglian, II late Anglian, III Hoxian, IV Present

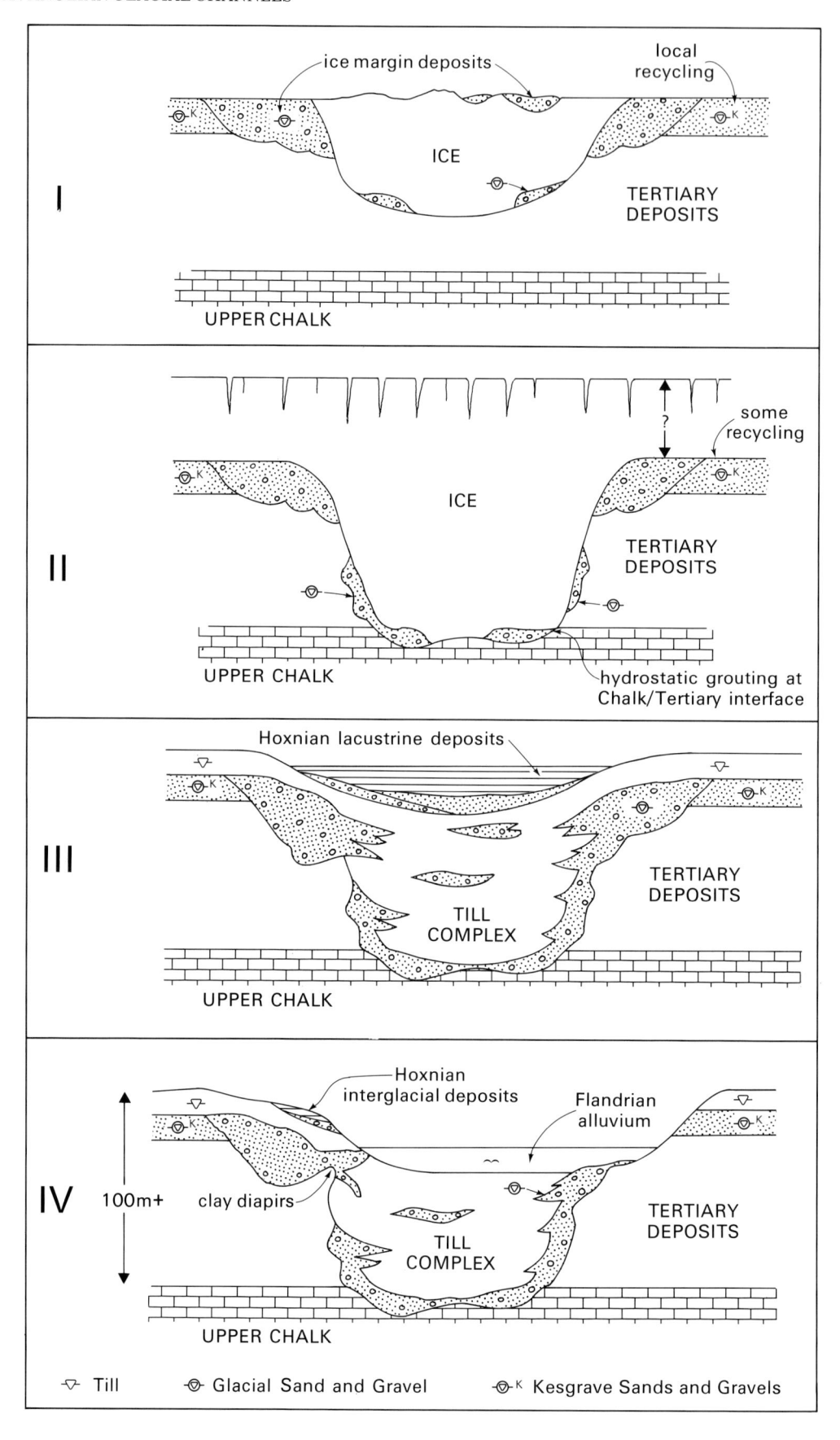

	Thickness m
HEAD Clay, sandy, buff to orange-brown, structureless, with scattered angular to subrounded flints; pocketed and cryoturbated base	1.7
GLACIAL SILT Silt and silty clay, buff and pale grey with orange mottling, finely laminated; laminae contorted and subvertical at about 2.8 m depth	1.7

Structureless silts were encountered in a comparable situation at Brand's Hill in a trial pit [5209 3766] where about 0.5 m of Head overlay 2 m of mottled silts.

A face in an old pit [5178 3707] at Neville Hill Wood, near Wendens Ambo, was cleared in 1980 and revealed the sequence:

	Thickness m
HEAD Silts, buff, friable, structureless, with scattered flint pebbles; relatively abundant chalk fragments below 4.4 m; poorly sorted medium gravel at 2.6 to 3.0 m; gravelly pockets at 3.6 to 3.85 m	4.75
Gravel, orange-brown, poorly sorted, with a sandy, silty matrix; dominantly flinty; flints angular to rounded; scattered sandstone and quartz pebbles; small irregular stringers of buff silt; chalky in lower part; base gradational	0.45
Silt, buff and grey-brown, with scattered flint chips and some layers of poorly sorted fine sand	0.2
Gravel, medium, orange-brown, poorly sorted, dominantly flinty; forms an irregular lens	0.1
GLACIAL SILT Silt, buff, weakly laminated	0.2
GLACIAL GRAVEL Gravel, medium- to fine-grained, poorly sorted	0.07
GLACIAL SILT Silt, weakly laminated, with scattered small gravelly pockets	0.13

A well [5163 3645] sited above the buried channel at Wendens Ambo encountered dominantly sands and gravels to 16.8 m, 'loams' with subordinate clays and chalky clays to 84.4 m, boulder clay to 90.2 m, overlying chalk with flints (?Middle Chalk).

To the north-east of Shortgrove, three closely spaced trial pits were dug near the present margin of the glacial deposits. The first [5302 3577] revealed:

	Thickness m
TILL Clay, chalky, slightly silty, with small sandy and gravelly pockets in basal 0.5 m	1.8
GLACIAL SAND AND GRAVEL Gravel, fine, clayey, sandy, with small flint, chalk and rare ironstone pebbles; slight dip to west	0.2
TILL Clay, chalky	0.2

A pit 40 m to the east, i.e. nearer the Chalk crop, proved 1 m of glacial sand. A further pit [5309 3577], 30 m to the east, showed:

	Thickness m
Top soil: stony loam	0.7
HEAD Sand, medium-grained, clayey, silty, orange-brown, homogeneous, with small angular flint chips locally; flinty, gravelly, with some chalk pebbles in basal 0.5 m; sharp base	0.8
GLACIAL SAND Sand, yellow-buff, medium- to fine-grained, poorly sorted, slightly chalky, with some scattered angular flint chips; no obvious bedding; discontinuous fine gravel lenses in basal 0.5 m (subangular to subrounded flint pebbles); olive and pale grey silty clay intraclasts up to 0.12 m long, subhorizontally aligned in basal 0.5 m; sharp, horizontal base	0.55
GLACIAL SILT Silt, clayey to silty clay, pale grey to yellow-buff, laminated; sharp base	0.31
GLACIAL SAND Sand, medium-grained, grey-buff, moderately sorted, chalky; no bedding structures seen; small ?flint chips	0.44

A trial pit [5315 3535] on the outcrop of Glacial Sand and Gravel to the south revealed 2.7 m of structureless, gravelly, sandy clay. The gravel component was largely of nodular, angular and rounded flints up to 0.2 m in diameter, with some exotic sandstone and granite pebbles. The flints were locally concentrated in stringers.

In a small disused gravel pit [5184 3442] 200 m west-north-west of Newport Grammar School, the following section was recorded:

	Thickness m
Top soil	0.4
GLACIAL SAND AND GRAVEL Sand, medium-grained, chalky, buff, well-sorted and laminated	0.4
Gravel, medium, well-sorted, flinty and chalky	2.0
Sand, as above	1.5
Obscured	2.5

The south-western face of an old gravel pit [5119 3416], 870 m east of Newport church, was cleared in 1980 and showed the following section:

	Thickness m
Fill	about 1.7
TILL Chalky boulder clay	0 to 0.25
GLACIAL SAND AND GRAVEL Gravel, fine-grained, moderately to well-sorted; orange, clayey, coarse sand matrix; clasts of white quartzite, ?igneous rock, angular weathered flints and vein quartz; the non-flint material was dominantly rounded to subrounded; crude lamination, inclined at about 40°N, defined by more sandy layers	up to 0.8
Gravel, medium- to coarse-grained, poorly sorted; dark brown sandy clay matrix; angular flints and scattered large angular flints; pockets of till (rafted?)	0.15

Gravel, fine-grained (as topmost unit); lensing out laterally to south-east and intersecting a mass of chalky boulder clay (injected and deformed ice-raft?); erosional base — 0 to 1.0

Gravel, coarse-grained, poorly sorted, dark brown, with battered angular and nodular flints, and small rounded vein quartz and quartzite pebbles; clay galls (decalcified boulder clay?); coarse sandy clay and clayey sand matrix — up to 1.0

The excavation continued below the pit-floor in coarse, loosely consolidated gravels as above, with irregular slightly chalky sand lenticles for 1.1 m. These rested with an irregular base on soft, putty chalk.

An excavation [5358 3399] in the base of an old gravel pit, near Debden Water, revealed the following glacial deposits (see also p.30):

Thickness m

GLACIAL SAND AND GRAVEL (MAINLY)
Gravel, coarse-grained, poorly sorted, with large nodular, rounded and angular flints; orange-brown, coarse-grained, sandy matrix; appreciable rounded pebble content; gravels become finer grained downwards below 0.55 m; irregular base — about 0.95

Clay, sandy, gravelly, brown, with large angular to rounded flints (solifluction deposit) — 0.15

Silts, buff to chocolate brown, structureless; irregularly interbedded with fairly well-sorted, buff, medium-grained sands; basal pebbly deposit comprising flints and angular chalk fragments — 0.5

Chalk breccia: angular chalk fragments in chalk matrix; scattered angular and rounded (patinated) flints; lenticular — 0.2

Gravels, as above — 0.5
Chalk breccia, as above — 0.5
Gravels, as above — —

A trial pit [5159 3369] near Frambury Lane, to the west of Newport, proved 2.4 m of Glacial Silt. These silts were buff to pale grey, finely and rhythmically laminated, with scattered fine sand lenticles and partings; localised small cross-bedded units were seen. Fine gravelly partings occurred below 1.35 m, although scattered small rounded flints and chalk pellets were present throughout the sequence. The bedding dipped south-east.

Four trial pits were sunk near Bonhunt Water to determine the stratigraphy of the drift. Three of these [5108 3323; 5109 3325; 5118 3333] proved chalky till to depths of 1.7, 3.2 and 2.2 m respectively. In the second pit the clay was moderately soft, with subhorizontal chalk-rich zones. In the third pit the clasts tended to be aligned horizontally, perhaps indicative of internal shearing. The fourth pit [5113 3332] proved:

Thickness m

Top soil and Head: brown silty clay — 1.0

TILL
Chalky boulder clay: wedges out to the north; gradational base — about 0.4

GLACIAL SILT:
Clay, silty, moderately firm, pale to medium grey with buff and medium brown mottling; scattered small chalk fragments; race in upper part; locally laminated, but crumpled and probably extremely disturbed — 1.3

Thus the till overlies disturbed Glacial Silt in the north but field mapping indicates that Chalk crops out directly beneath the till south-westwards.

Glacial Sand and Gravel of varied thickness and depth within the buried channel complex was proved in numerous trial boreholes for the M11 motorway south and west of Newport; one of the thickest gravel deposits was recorded in a borehole [5124 3356] near Bonhunt:

	Thickness m	*Depth* m
Soil	1.1	1.1
TILL		
Clay, silty, sandy, pale greyish brown, firm, with stones and chalk fragments	2.0	3.1
GLACIAL SAND AND GRAVEL		
Sand and gravel; with flint fragments and some yellowish grey silty clay	3.5	6.6
?GLACIAL SILT		
Clay, silty, stiff, brownish grey	3.0	9.6
GLACIAL SAND AND GRAVEL		
Sand and gravel: fine to coarse chalk gravel and grey clayey sand with flints in parts	seen 20.9	30.5

In contrast, a nearby borehole [5121 3356] encountered mainly silts and clays:

	Thickness m	*Depth* m
HEAD	0.9	0.9
TILL		
Flints, stones and chalk fragments in brown sandy clay	2.2	3.1
GLACIAL SAND AND GRAVEL		
Flints, gravel and clayey sand	1.2	4.3
Clay, silty, firm, greyish brown	0.3	4.6
Sand, coarse- to fine-grained, brown, with gravel, flints and a little clay	8.5	13.1
GLACIAL SILT		
Clay, silty, stiff, greyish brown	5.1	18.2
Clay, silty, grey, very stiff	3.6	21.8
Silt, clayey, grey, very stiff, with beds of fine-grained sand and scattered chalk particles	8.7	30.5

A section with laminated clays was recorded at [5164 3320] in a trackside cutting adjacent to the M11 motorway.

	Thickness m
GLACIAL SAND AND GRAVEL	
Gravel, chalky, flinty, fairly well-sorted	1.0
GLACIAL SILT	
Clay, silty, bluish grey and brown, with well-defined fine lamination	1.5

To the south of Newport, on the flanks of the Cam valley, several outcrops of Glacial Silt were investigated by trenching. To the west of Newport Pond a trial pit [5187 3290] exposed 2.6 m of buff, generally well-laminated, friable silts with partings of fine-grained sand and very silty clay, beneath 0.4 m of top soil. The silts were structureless in the top 1.0 m. Although much of the bedding was planar and inclined to the south-east, cross-bedding was seen in the lowest beds. A further 0.5 m of silt was augered in the base of the pit.

Further south a trial pit [5200 3253] showed:

	Thickness m
Top soil and wash (from till): sandy clay, dark brown, with scattered angular flints	0.9
GLACIAL SILT	
Silt, buff, friable, compact, homogeneous, with irregular pale grey clayey lenticles; discernible brecciated structure towards base; rare scattered angular flints	1.1
Silt, buff and pale grey, with local sand units, locally clayey; well-developed small scale cross-bedding, contorted in places; clayey, chalky (1 cm) silt lenticle at base; rare dropstones	0.45
Silt, buff and pale grey, with small scale cross-bedding; local clay lenticles especially at 2.6 m depth; abundant penecontemporaneous fold structures towards top of unit; some penecontemporaneous (?thrust) faulting; rare dropstones	0.15
Silt, slightly clayey, fairly homogeneous, with local sandy intercalations; bedding and ?deformation structures poorly discernible	0.8

Calcareous 'race' was abundant in the more clayey zones.

In contrast, a pit 30 m to the west [5197 3252] showed:

	Thickness m
Top soil: silty clay, brown, with small scattered flints	0.4
GLACIAL SILT	
Clay, pale bluish grey with brown mottling, structureless at top; small chalk fragments variably disposed; faintly laminated below 0.9 m depth, with few chalk fragments; well-laminated with thin medium sand partings below 1.3 m; interbedded with silt and fine-grained yellow sand to indistinct base	1.2
GLACIAL SAND AND GRAVEL	
Gravel, poorly sorted, coarse, slightly clayey, with large, nodular and angular flints in coarse clayey sand matrix; sand lenticles near base	0.5

An excavation [5241 3203] further south in another outcrop of Glacial Silt showed:

	Thickness m
Loam, dark brown, silty, with scattered flints	0.3
HEAD (reworked Glacial Silt)	
Silt, clayey, structureless, buff, with scattered chalk pellets, angular flint pebbles and clay galls; sharp base	0.6
Silt, clayey, sandy and gravelly, buff to orange-brown, variegated, with sand pockets and abundant angular flints; sharp base (base of soliflucted material)	0.3

	Thickness m
GLACIAL SILT	
Silt, fine-grained, sandy, locally clayey, buff, locally pale grey, well-laminated throughout, compact, with ?load structures; local small-scale cross-bedding; silty clay bed with some race from 1.65 to 1.70 m; silts becoming coarser downwards, with local pebble stringers; gradational base	0.9
Sand, medium- to coarse-grained, buff to grey, poorly sorted, chalky; cross-bedded foresets dip to the south-west; penecontemporaneous faulting; finer-grained towards sharp base	0.2
Silt, buff, wispy bedded, locally well-laminated, compact	0.5

The complex alternation of Glacial Sand and Gravel with other drift deposits at the margins of the buried channel is well illustrated by a trial borehole for the M11 [5221 3174]:

	Thickness m	Depth m
Top soil	0.3	0.3
TILL		
Clay, sandy, brown, firm, with chalk fragments	1.1	1.4
GLACIAL SAND AND GRAVEL		
Sand, clayey, orange-brown, with scattered pebbles	2.9	4.3
TILL		
Clay, silty, orange-brown, firm, with large flints and some sand	1.2	5.5
Clay, silty, yellow, firm, with stones and chalk fragments	1.8	7.3
GLACIAL SAND AND GRAVEL		
Sand, medium-grained, clayey, orange-brown, and silty sand, with scattered chalk fragments	2.5	9.8
Sand, clayey, brown, with putty chalk	0.6	10.4
Sand, coarse-grained, clayey, orange-brown, and sandy clay with stones and chalk fragments	1.2	11.6
UPPER CHALK	1.5	13.1

Soliflucted till deposits, shown as Head on the published map, were encountered in a trial pit [5114 3086] near Howells:

	Thickness m
HEAD	
Clay, silty, medium-brown, firm, with scattered angular flints; siltier in lower part; coarse angular flints in 0.15 m-thick band at base (solifluction lobe)	1.25
Clay, silty, brown, firm, with scattered small angular flints; siltier in lower part; abundant small shattered flints with a few large nodular flints concentrated in 0.25 m-thick band at base (solifluction lobe)	0.80
Clay, slightly silty, brown, softer than above, with scattered small angular flints (?brecciated)	seen 0.75
Clay, slightly silty, smooth, ?flinty	augered to 1.00

A nearby trial pit [5118 3083] demonstrated that chalky till underlies the shallow tributary valley:

HEAD
Clay, silty, brown, but buff and more clayey in parts; abundant large angular, nodular and rounded flints; irregular base — about 1.15

Clay, pale to medium-grey, brown-mottled or red-stained in parts, with scattered sand grains and small angular flints; irregular ?transitional base — 0.35

TILL
Clay, silty, pale to medium-grey, with abundant chalk fragments and small weathered flint chips; indistinct ?flow-banding; weak clast alignment; open earthy texture — 0.5

On the east side of the Cam valley, in London Jock Wood, a trial pit [5263 3075] exposed a sequence with laminated silts as follows:

Thickness m

HEAD
Clay, silty, dark brown, with scattered angular and rounded patinated flints; paler near base, with more abundant chalk fragments; pocketed base — 1.0 to 1.2

?GLACIAL SILT
Silts, buff, orange-brown and pale grey, finely interlaminated in parts with orange and grey, fine-grained sands; local lenses of sand; laminae generally horizontal but showing local disturbance and injection; chalk cobbles (up to 0.2 m in diameter) at 2.25 m depth with larger weathered chalk fragments below; thin ?carbonaceous laminae developed locally in silty clays above irregular base — up to 1.9

UPPER CHALK
—

The field relationships in this area clearly indicate that the glacial sequence occupies a channel along the line of the Cam valley.

To the east and south-east of The Hall, Ugley, three trial pits were dug in glacial deposits. One pit [5238 2861] showed:

Thickness m

Top soil and silty Head, ochreous and pale grey-mottled — 1.1

GLACIAL SILT
Clay, silty, sandy, grading to clayey silt, brown, orange and pale grey mottled, generally structureless; sandier intercalations seeping water; blebs of derived Tertiary glauconitic material; soft, friable, with traces of relict lamination (probably destructured) — 1.7

Farther south, near Ugley Park, a trial pit [5280 2761] encountered similar material:

Thickness m

Top soil and flinty clayey wash — 1.2
Clay, very silty, soft, brown and grey mottled, ?carbonaceous (?old ditch-fill) — 0.15

GLACIAL SILT
Clay, very silty, slightly sandy, mottled grey, buff and orange, structureless, with few scattered angular flints; irregular bodies of more sandy and more clayey material occur as diffuse intercalations (all poorly sorted); irregular base inclined to east — 1.35

GLACIAL SAND AND GRAVEL
Sand, medium-grained, moderately well-sorted, orange-grey with irregular brown staining; no bedding seen; small pale grey clay galls, vertical at one point (seen in west face only) — 0.6

The glacial sands which overlie the structureless clay were encountered in a pit [5260 2760] upslope:

Thickness m

Top soil and flinty fine-grained sandy Head, ochreous and pale grey-mottled — 0.6

GLACIAL SAND AND GRAVEL
Sands, fine- to medium-grained, clayey, buff and pale grey-mottled; pale grey clay gall (0.2 m diameter) at 1.1 m depth; slightly clayey, coarse-grained sand below, with abundant fine, rounded to subangular gravel (less than 15 per cent quartz); (?solifluated unit); sharp base at — 0.75

Sand, fine- to medium-grained, buff, locally iron-stained, unbedded; water seepage below 1.8 m depth; coarse rounded cobbles at about 2.1 m depth; medium rounded to subangular gravelly sands below — 1.25

Elsenham and Bishop's Stortford area

The Ugley Sand Quarry [517 277] exposed up to 10 m of glacial sands and coarse flinty gravels, overlain by chalky till which was reputedly up to 12 m thick in the north-east face. The gravels were poorly sorted and interbedded with white, chalky, fine- to medium-grained sands. Planar bedding, complex channel-fill structures and large-scale cross-beds were visible. The gravels also contained discrete grey clay or clayey silt partings and intraformational ice-wedges. Where the till cover was thin, the gravels showed decalcification structures.

In the north-east face of the pit, the till contained intercalations of silts with laminated beds, and irregular bodies of silty sands with complex convoluted structures and lenticles of firm olive-grey clay (Figure 7). In this part of the pit the bedrock rose appreciably, and soft clays derived from the bedrock were injected into the overlying sediments.

The section above the trench described (on p.12) in Tertiary deposits [5206 2792] was recorded as follows:

Thickness m

Top soil — 0.2

TILL
Clay, compact, stiff, olive-grey, chalky; locally (?flow) laminated at base (0.5–0.7 m thick); large flints common at base; sharp ?channelled base at south end of section — 0 to 4.5

GLACIAL SAND AND GRAVEL
Sands, medium- to coarse-grained, fairly well-sorted, buff, yellow and grey with reddish brown mottling; cross-bedded units and fine lamination

locally; scattered thin clay laminae; compact clayey silty sand with thin coarse-grained flint gravel (containing scattered quartzes) at base; this unit thickens and channels downwards at south end of section   0.3 to 2.6

A gravel pit [490 237] at Hazelend, near the Bishop's Stortford bypass, exposed 4 m of planar-bedded chalky Glacial Sand and Gravel beneath 2 m of chalky boulder clay. Fine to medium chalk detritus was more common in the upper part of the gravel.

The spread of Glacial Sand and Gravel near Hazelend [496 244] is probably thin (less than 4 m thick) and locally degraded. In the western part of the outcrop the gravels overlie the boulder clay, but to the east they rest directly on the Upper Chalk. They form a broad flattish feature which falls gently southwards to the buried channel of the River Stort.

Glacial silts near Foxdells [488 228] apparently have a steeply inclined contact with the boulder clay to the north. This relationship may be an ice-marginal feature indicating that the silts approximately define the north-western limit of the channel. A nearby degraded gravel pit [489 228] was apparently dug in up to 6 m of moderately sorted rounded and angular flint gravels with intercalations of chalky sand. The presence of boulder clay spoil suggests that there may be interbeds of till within the sequence.

Boreholes for the Bishop's Stortford Bypass showed that the buried channel (see Figure 12c) is largely infilled with gravel along this traverse. However, a nearby borehole [4942 2299], some 400 m to the south-west, proved the following sequence:

	Thickness m	*Depth* m
HEAD		
Gravel, clayey, brown, passing down to coarse gravel	3.0	3.0
Clay, brown, slightly micaceous, with some chalk pebbles	1.4	4.4
Gravel, fine, flinty	0.8	5.2
LACUSTRINE DEPOSITS		
Clayey silts and silty clays, greenish grey; flint fragments locally; *Chara* marl with freshwater ostracods, bivalves and plant debris from 7.6 to 10.1 m depth	4.9	10.1
Clay, green and brown, with medium to coarse flint gravel (?solifucted unit)	3.0	13.1
GLACIAL SAND AND GRAVEL		
Gravel, fine, in coarse sand matrix	6.4	19.5
TILL		
Clay, sandy, grey, with chalk pellets and angular flints; local intercalations of gravel?	7.0	26.5
GLACIAL SAND AND GRAVEL		
Chalk breccia in sandy silt matrix	0.9	27.4
Gravel, chalky, flinty	0.9	28.3
UPPER CHALK	5.2	33.5

Other boreholes proving the buried channel are listed in Table 4.

A temporary excavation [4925 2129] near Hockerill Street, Bishop's Stortford, showed the following sequence:

	Thickness m
Fill and sandy wash	up to 1.4
GLACIAL SAND AND GRAVEL	
Gravels, medium, flinty, poorly bedded, moderately sorted, dominantly rounded; lenses of chalky sand	2.5
TILL	
Chalky boulder clay	—

This locality is probably close to the eastern margin of the channel.

A former gravel pit [4894 2022] at the football ground exposed:

	Thickness m
TILL	
Chalky boulder clay (draped in part)	?2.0
GLACIAL SAND AND GRAVEL	
Sandy clays and clayey sands, pebbly, brown and pale grey mottled	1.0
Sands, coarse-grained, slightly clayey, orange, with rounded and angular flint pebbles	2.0

SEVEN

Quaternary: Hoxnian to Flandrian

The deposits which postdate the Anglian glaciation have a restricted distribution and are by no means fully representative of the time interval. Lacustrine deposits in the Cam valley are of Hoxnian age, whereas the Head deposits and river terrace sands and gravels are probably largely Devensian. The alluvium in the major river valleys was deposited in the Flandrian stage; similar clayey material in the minor valleys near Bonhunt [512 334] was probably formed in late Devensian to Flandrian times.

LACUSTRINE DEPOSITS

Organic silts which revealed a Hoxnian pollen spectrum were recognised near North Hall by Baker (1976). Trial boreholes demonstrated that these lacustrine deposits overlie till at a minimum level of 56 m above OD. A BGS trial-pit [5243 3031] proved shelly bioturbated silts and clays resting on till at about 69 m above OD. These organic deposits appear to be local and probably accumulated in hollows resulting from the melt-out of buried ice in a glacial channel, prior to an episode of deposition during Lowestoft Lateglacial to Hoxnian subzone III b times (Baker, 1976, p.288).

A borehole [4942 2299] near the Bishop's Stortford by-pass proved 4.9 m of silts and clays, with shelly *Chara* marls in the lower half, beneath 5.2 m of flinty clays and gravels (Head). The marls overlie glacial deposits which rest on Upper Chalk at 28.3 m depth (p.39). These shelly sediments apparently occupy a hollow similar to that at North Hall, but have been preserved beneath later solifluction deposits. Comparable organic sediments were not recognised in the trial boreholes for the by-pass, but shelly organic clays were recorded beneath terrace deposits in Bishop's Stortford (p.32).

A small outcrop of lacustrine deposits has been mapped near Morrice Green [415 354]. They comprise lilac and grey clays which are possibly ?organic or humic-rich and which were described by Whitaker et al. (1878, p.64) as 'mottled loam', at least 1.5 m thick. These sediments may occupy a former kettle-hole analogous to those described near Hatfield (Sparks et al., 1969).

HEAD AND HEAD GRAVEL

Head deposits are present on the lower slopes of the valleys. They typically comprise silts and sandy clays with flinty intercalations, particularly near the base. They formed by downslope wash and solifluction under periglacial conditions, and their composition reflects the parent materials upslope; thus, in the vicinity of Chalk outcrops, Head deposits commonly include reconstituted chalky sludges and breccias. Locally the Head tends to be preferentially preserved on north- and east-facing slopes.

Head Gravel occurs extensively in the Ash valley near Little Hadham (see p.5). It has also been distinguished near Widdington (see below) and near Latchmore Bank [498 187]; at the latter locality it consists dominantly of clayey, coarse-and fine-grained, angular, flint gravels. These gravels may be related to a 'train' of such deposits which extends south-south-eastwards on the adjoining district through Hatfield Heath and Matching Green (Millward et al., 1987). These deposits were interpreted as fluvioglacial (supraglacial) in origin but have suffered the effects of subsequent cryoturbation and possibly solifluction.

RIVER TERRACE DEPOSITS

A low-level suite of River Terrace Deposits (Terrace 1) has been recognised in the Ash, Stort, Chelmer and Pant valleys. These sandy and clayey flint gravels are patchily distributed adjacent to, and up to 3 m above the floodplains. A borehole [5015 2431] near Limekiln Lane, Stansted Mountfitchet, proved 5.5 m of gravels and carbonaceous silts; the silts form an interbed, 2.3 m thick. Up to 3.7 m of sandy and clayey gravel were recorded [6291 2312] in the River Chelmer valley near Great Dunmow.

A higher terrace (Terrace 2) of similar composition occurs in the valleys of the Rivers Ash and Stort near the southern margin of the district, up to 4.5 m above the floodplains.

ALLUVIUM: MERE DEPOSITS

Three discontinuous alluvial tracts in the upper part of the River Cam valley were studied by Baker (1976) and identified as distinct basins marking areas of former drainage impedence. These areas (Bonhunt Water [516 342], Newport Pond [523 332] and Quendon Want [524 314]) were shown to be former mere-like lakes, impounded behind gravelly solifluction flows. Other examples probably include those at Debden Water and near Wicken Bonhunt. On the published geological map the sediments downstream from these tracts of alluvium are depicted mainly as Head, although Head Gravel has been distinguished in one area [525 325] northwest of Widdington. Baker (1976, p.302) thought that the gravels 'interdigitate with the diamicton deposits whose structural and textural characteristics indicate debris sliding and earthflow reworking of the plateau till edge' (but see p.28).

The alluvial sediments comprise clays, silts, sands and fen peats up to 7.5 m thick at Bonhunt Water (Baker, 1976); silty clays are common at the surface. The pollen stratigraphy indicates the presence of the Late Glacial zones III-IV transition to Sub-Atlantic Zone VIII; the earlier dating is confirmed by a radiocarbon assay of 10 040 ± 160 BP (Baker, 1976, p.302).

ALLUVIUM: FLOODPLAIN DEPOSITS

Alluvium occurs beneath the floodplains of all the major river valleys of the district and, to a lesser extent, in their tributaries. It consists of soft to firm, pale grey and brown silts and silty clays with some organic detritus. Thin gravel beds commonly occur near the margins and at the base of the deposit. Typical thicknesses range from 1.5 to 4 m. Exceptionally, in the Pant valley near Great Sampford, up to 15 m were proved [6400 3569], of which the lower 13.5 m were detrital peats. This increased thickness may be due to solution of the Chalk beneath.

In the River Chelmer valley near Great Dunmow, thick suballuvial gravels were proved above the London Clay, and were described as Undifferentiated River Terrace Deposits by Thomas (1982). These angular flint gravels, which contain beds of detrital peats and sandy and clayey silts, range in thickness from 6.8 m near Dunmowpark [6377 2176] to 14.7 m near Butcher's Pasture [6107 2439]. The origin of these gravels is uncertain, but those which underlie the alluvium of the Cam and Stort valleys are mainly related to the glacial channels which underlie these tracts (see p.28).

CALCAREOUS TUFA

Calcareous tufa occurs as a spring-line deposit near Martels [638 197]. This is a soft, pale grey silt-grade deposit formed of calcium carbonate, which was precipitated from waters percolating from the boulder clay.

DETAILS

A trial pit [5243 3031] near North Hall exposed the following:

	Thickness m
HEAD	
Silt, clayey, dark brown, with scattered angular flints; moderately friable	0.9
Silt, clayey, buff to pale grey, with abundant comminuted thin shell fragments; no obvious internal structure; diffuse, but apparently channelled base; (?derived from below)	0.1 to 0.5
LACUSTRINE DEPOSITS	
Silt, calcareous, pale grey, mottled in shades of pale olive-brown, with abundant comminuted shell fragments; apparent bioturbation, especially in upper part, shown by subvertical vermiform colour mottling in pale grey; passing down to	up to 0.9
Clay, silty, pale grey, with tufaceous laminae; gritty texture (race); shell fragments as above; ochreous and softer with increasing water content in lower part	0.6
TILL	
Clay, silty, chalky; dark bluish grey in lower part with distinctive faceted ice-scraped pebbles	0.5

A borehole [5238 3010] near North Hall proved soft sandy clays overlying stiff brown and grey silty clays to 6.1 m. These were shown to overlie till at 6.6 m depth in a borehole [5237 3029] to the north. The silty clays may be comparable to the lacustrine deposits described above, but a trial pit [5229 3008] nearby proved only 1.5 m of sandy, flinty clay (Head) on chalk.

The Upper Chalk in the Neville Hill area, Wendens Ambo, is extensively covered by a wash of grey and brown sandy clays. A trial pit [5153 3713] proved 2.9 m of this material, possibly of increased thickness as a result of dissolution of the bedrock:

	Thickness m
HEAD AND ?SOLUTION-HOLLOW FILL	
Clay, slightly sandy, brown, with scattered flints (mostly angular) and chalk fragments	1.1
Clay, sandy, orange-brown, with scattered flints and partly decalcified chalky patches about 0.05 m thick	0.3
Silt, loamy, creamy white, with scattered flints; irregular but fairly sharp top and base with irregular bands and pockets of orange-brown sandy clay; becomes streaky or laminated with flint and chalk detritus in places	0.2
Clay, sandy, orange-brown, with scattered angular flints and some irregular sandy stringers; sharp base	0.95
Marl, silty, sandy, buff, with rounded chalk pebbles and scattered flints and sandy patches; chalk pebbles commonly show solution pits where in contact; base very irregular	0.35
UPPER CHALK	
Chalk breccia: fragments of chalk in hard brecciated calcite matrix	0.1

Downslope, to the east, another pit [5172 3705] penetrated only 1.0 m of sandy flinty clay, resting with a very irregular base on Upper Chalk.

In Brakey Ley Wood, near Shortgrove Hall, a trial pit [5337 3580] revealed at least 2.2 m of sandy clay and gravelly loam (Head). A lower gravelly unit, a middle chalky silty unit and an upper brown sandy clayey unit were distinguished.

Excavations [about 530 366] for a culvert by Joshua's Bridge, near Wendens Ambo, showed the maximum recorded thickness of 4.9 m of alluvium. The complete section was: top soil to 0.3 m, clay to 1.2 m, resting on peat to 4.9 m, on gravel, touched. The section was highly fossiliferous, two cartloads of mammal bones being removed from the base of the peat (information from Saffron Walden Museum).

EIGHT
Economic geology

BUILDING MATERIALS

The Tertiary deposits have been used locally for brickmaking. The Woolwich and Reading Beds, in particular, were exploited in brickyards in the Bishop's Stortford area and the relevant sections are described in Chapter 3.

SAND AND GRAVEL

Surveys of the sand and gravel resources of this district were undertaken by the former IGS Industrial Minerals Assessment Unit between 1975 and 1981. The results of these studies were published as detailed on p. iv.

The physical criteria used to define sand and gravel as a resource were:
a) the deposit should average at least one metre in thickness,
b) the ratio of overburden to sand and gravel should be no more than 3:1,
c) the proportion of fines should not exceed 40 per cent (the fines, that is clay and silt, comprise particles less than 0.063 mm mean diameter),
d) the deposit should lie within 25 m of the surface.

The major resources within the Great Dunmow district are the spreads of Red Crag, Kesgrave Sands and Gravels, and Glacial Sand and Gravel. Other deposits are of minor economic importance.

In 1988 there were six working sand and gravel pits in the district: at Hazelend [490 237], Hollow Lane, Widdington [530 310], Elsenham [546 268], Canfield [578 211], Armigers [597 299] and Cowlands Farm, Stebbing [669 233].

Summarising the results of the published surveys, which cover about 70 per cent of the district, it is estimated that total resources of sand and gravel in the areas assessed amount to about 930 million m^3; of these at least 80 million m^3 are classified as non-Kesgrave materials and are generally of a lower quality. Reservations applied to the estimates of resources are contained in the published reports.

The mean gravel component within the areas of sheets TL 52, TL 62 and TL 63 varies between 16 and 22 per cent by weight. Such values are locally significantly lower, and may be as little as 7 per cent in individual resource blocks in the northern part of the district, where the sandy facies of the Kesgrave Sands and Gravels predominates.

HYDROGEOLOGY AND WATER SUPPLY

The district is located mainly within Hydrometric Areas 37 and 38, with a small part also within Area 33. To the north and east, the water resources are administered by the Anglian Water Authority, and to the south and west by the Thames Water Authority. The management units involved (Monkhouse and Richards, 1982) are Anglian 12 and 46, and Thames 04, 05, 06 and 08.

Drainage of the southern part of the district is accomplished by the rivers Ash, Stort and Roding, which flow southwards into the River Thames. In the east, the rivers Pant and Chelmer flow east and then south to the Thames estuary. The north of the district is drained by tributaries of the River Cam. All these streams are moderately 'flashy', being characterised by sharp increases in flow after wet periods; low flows are commonly less than 10 per cent of the mean.

Rainfall tends to be a little heavier in the west of the district than in the east, but the annual average is of the order of 600 mm. The mean annual evaporation is about 450 mm.

The earliest work on water resources in this district was carried out by Whitaker and Thresh (1916) and by Whitaker (1921). Catalogues of wells were published by the Geological Survey in the postwar years (Woodland, 1945; Sayer and Harvey, 1965). Hydrological surveys were published by the Ministry of Housing and Local Government (Anon, 1960, 1961, 1962), while summaries of water resources and demands were prepared by the Thames Conservancy (1969), the Essex River Authority (1971) and the Lee Conservancy Catchment Board (1974).

The total licensed water take for the district is currently 16.65 million cubic metres per annum (m^3/a), of which 7 per cent is from surface water intakes and 93 per cent from groundwater. The latter is taken almost entirely from the Chalk, with less than 1 per cent coming from the superficial deposits (sands and gravels). All of the public water supply sources in the area use groundwater from the Chalk, the licensed total being 13.39 million m^3/a (included with the public supply totals in Table 5). The actual abstraction of groundwater during 1987 appears to have been of the order of half the licensed total. A borehole in the Chalk at Great Sampford [645 351], which is operated by the Anglian Water Authority to augment the flow of the River Pant, is licensed for 1.05 million m^3/a (and is included with the public supply totals in Table 5).

The Gault is an aquiclude and therefore is hydrogeologically important because it forms a lower limit to the major Chalk aquifer; it does not crop out within the district.

The Chalk is the major aquifer and underlies the entire district. It is present at outcrop in the north and west, although extensively covered by drift. To the south and east this aquifer is confined by Tertiary strata. Groundwater is taken mostly from the Upper and Middle Chalk, and very little from the less permeable Lower Chalk.

The groundwater in the Chalk is contained within and flows through fissures which tend to be distributed at random. Consequently, there is considerable variation in the yield of boreholes. The yield also varies in proportion to the

Table 5 Details of licensed water abstraction for 1988 (million cubic metres per annum) (based on information supplied by the Anglian and Thames water authorities)

Source	Public supply	Industry	Agriculture	Spray irrigation	Other	Totals
Surface	nil	nil	6.6	1 082.3	6.8	1 095.8
Chalk	13 389.3	1 831.7	94.4	129.6	18.6	15 463.7
Gravels	nil	1.7	66.7	18.2	nil	86.4
Totals	13 389.3	1 833.4	167.7	1 1230.1	25.4	16 645.9

borehole diameter, to the length of the borehole open to the saturated aquifer, and (in the aquifer outcrop) to the depth of the rest water level from the ground surface. A borehole of 300 mm diameter in the Chalk outcrop, open to 50 m thickness of saturated chalk, and with the rest water level at or near to the ground surface, may on average yield some 2000 million cubic metres per day (m^3/d) for a drawdown of 10 m. However, there would be approximately a 20 per cent chance of the yield being less than 1200 m^3/d for the same drawdown. For smaller diameters and penetration, the yield would be less; for a borehole in a similar situation but with a diameter of 150 mm and open to only 30 m of saturated chalk, the mean yield would be of the order of 950 m^3/d for 10 m drawdown and there would be a 20 per cent chance of the yield being less than 550 m^3/d.

Borehole yields are less under Tertiary cover, although variations in the thickness of this cover do not appear to be statistically significant over this district. The mean yield of a 300 mm diameter borehole open to 50 m thickness of saturated chalk is about 1300 m^3/d for a 10 m drawdown; in this case, there would be approximately a 20 per cent chance that the yield would be less than 450 m^3/d.

The groundwater in the outcrop of the Chalk is typically of the calcium bicarbonate type (see 1, 2 and 3 of Table 6). Where drift cover is thin or absent, nitrate may be present in concentrations exceeding 1.0 milligrammes per litre (mg/l) [as N]. Beneath more extensive, impermeable drift (such as Boulder Clay), the concentration of nitrate is normally less than 1.0 mg/l. Farther downdip, where the Chalk passes beneath Tertiary cover, the groundwater type changes to a calcium-sodium-bicarbonate type (4 of Table 6), indicating a softening effect due to ion exchange of calcium for sodium, together with a reduction of the sulphate ion concentration. In the south-east corner of the district, where the Tertiary cover is thickest, the groundwater changes to a sodium-bicarbonate-chloride type, the already softened water becoming mixed with saline connate water (5 of Table 6).

The Thanet Beds are of little importance as a groundwater resource, the thickness being small and the permeability low. The Woolwich and Reading Beds would probably yield some water from the sandy facies of the 'Bottom Bed', although supplies might be rather variable across the district. The Reading facies forms an aquiclude yielding little or no water. The London Clay is also an aquiclude, although very small yields of poor quality water (generally rich in iron and sulphate) can sometimes be obtained from shallow shafts excavated in the weathered zone near to the ground surface.

Where present in sufficient extent and thickness, groundwater can be taken from boreholes in the Crag. However, this water is often very hard and commonly rich in iron, which renders it rather undesirable even for small supplies.

Shallow shafts and boreholes yield agricultural and domestic supplies from the superficial deposits, mainly from the Kesgrave Sands and Gravels, and from glacial gravels, although yields from individual sources are limited to a few hundred cubic metres per day at best. Sources in these deposits occasionally yield iron-rich water, and they are very vulnerable to surface pollution; the concentration of nitrate, for example, may locally exceed 12 mg/l [as N]. Sources in the Till rarely yield more than seepages which are likely to be insufficient even for a domestic supply.

Table 6 Typical groundwater analyses from the Chalk (content in milligrams per litre). (Based on information received from the Anglian and Thames water authorities)

	1	2	3	4	5
Total dissolved solids	414	400	570	490	960
Carbonate hardness (as $CaCO_3$)	304	270	290	300	160
Total hardness (as $CaCO_3$)	344	280	420	325	160
Calcium (as Ca)	127	98	140	95	33
Magnesium (as Mg)	6.2	9.5	19.0	21.0	19.0
Sodium (as Na)	13.5	16.0	22.0	33.0	290
Potassium (as K)	1.9	3.0	4.5	13.0	19.0
Bicarbonate (as HCO_3)	370	329	353	365	353
Sulphate (as SO_4)	23	16	105	51	95
Chloride (as Cl)	22	24	30	32	270
Nitrate (as N)	4.5	2.1	<1.0	<1.0	<1.0
Iron (as Fe)	<0.01	0.19	0.11	—	0.05
Fluoride (as F)	0.2	0.36	0.70	0.05	1.80
pH	6.9	7.3	7.2	7.8	8.0

1 Calcium bicarbonate water with nitrate <1.0 mg/l, indicative of aquifer outcrop without drift cover, or with thin and patchy drift. Located in the north-west of the district.
2 As above, but further to the south-east.
3 Calcium bicarbonate water with nitrate <1.0 mg/l; drift cover more extensive and less permeable.
4 Calcium/sodium bicarbonate water; Chalk passing down beneath the Tertiary cover.
5 Sodium bicarbonate/chloride water; Chalk under thick Tertiary cover in the south-east corner of the district.

REFERENCES

Most of the references listed below are held in the Library of the British Geological Survey at Keyworth, Nottingham. Copies of the references can be purchased subject to the current copyright legislation.

ALLSOP, J M, and SMITH, N J P. 1988. The deep geology of Essex. *Proceedings of the Geologists' Association*, Vol. 99, 249–260.

ANON 1960. River Great Ouse Basin Hydrological Survey, Hydrometric Area 33. 99 pp. (London: HMSO for Ministry of Housing and Local Government.)

— 1961. Essex Rivers and Stour Hydrological Survey, Hydrometric Areas 36 and 37. 123 pp.(London: HMSO for Ministry of Housing and Local Government.)

— 1962. River Lee Basin Hydrological Survey, Hydrometric Area 38. 53 pp. (London: HMSO for Ministry of Housing and Local Government.)

BADEN-POWELL, D F W. 1948. The Chalky Boulder Clays of Norfolk and Suffolk. *Geological Magazine*, Vol. 85, 279–296.

BAKER, C A. 1976. Late Devensian periglacial phenomena in the upper Cam valley, north Essex. *Proceedings of the Geologists' Association*, Vol. 87, 285–306.

— 1977. Quaternary stratigraphy and environments in the upper Cam valley, north Essex. PhD thesis, University of London.

— and JONES, D K C. 1980. Glaciation of the London Basin and its influence on the drainage pattern: a review and appraisal. Chapter 6 in *The shaping of southern England*. JONES, D K C (editor). (Academic Press.)

BATEMAN, R M, and MOFFAT, A J. 1987. Petrography of the Woolwich and Reading Formation (Late Palaeocene) of the Chiltern Hills, southern England. *Tertiary Research*, Vol. 8, 75–103.

BERGGREN, W A, KENT, D V, and FLYNN, J J. 1985. Palaeogene geochronology and chronostratigraphy. 141–195 in Geochronology of the geological record. SNELLING, N J (editor). *Memoir of the Geological Society of London*, No. 10.

BOSWELL, P G H. 1914. On the age of the Suffolk valleys and the buried channels of drift. *Quarterly Journal of the Geological Society of London*, Vol. 69, 581–620.

BRISTOW, C R. 1985. The geology of the country around Chelmsford. *Memoir of the British Geological Survey*, Sheet 241 (England and Wales).

— and COX, F C. 1973. The Gipping Till: a reappraisal of East Anglian glacial stratigraphy. *Quarterly Journal of the Geological Society of London*, Vol. 129, 1–37.

BUTLER, D E. 1981. Marine faunas from concealed Devonian rocks of southern England and their reflection of the Frasnian transgression. *Geological Magazine*, Vol. 118, 679–697.

CATT, J A. 1977. 96–97 in discussion of Middle Pleistocene stratigraphy in south-east Suffolk. ROSE, J, and ALLEN, P. 1977. *Quarterly Journal of the Geological Society of London*, Vol. 133, 83–103. In Proceedings of the Geological Society of London, Vol. 134.

CHARLESWORTH, J K. 1957. *The Quaternary Era with special reference to its glaciation.* 2 vols. (London: Edward Arnold.)

CLAYTON, K M. 1957. Some aspects of the glacial deposits of Essex. *Proceedings of the Geologists' Association*, Vol. 68, 1–21.

— and BROWN, J C. 1958. The glacial deposits around Hertford. *Proceedings of the Geologists' Association,* Vol. 69, 103–119.

COSTA, L I, and DOWNIE, C. 1976. The distribution of the dinoflagellate *Wetzeliella* in the Palaeogene of north-western Europe. *Palaeontology,* Vol. 19, 591–614.

COX, F C. 1985. The tunnel-valleys of Norfolk, East Anglia. *Proceedings of the Geologists' Association,* Vol. 96, 357–369.

CURRY, D. 1958. *Lexique Stratigraphique International*, Vol. 1, Part3a XII, *Great Britain Palaeogene*. (Paris: Centre National Recherche Scientifique.)

— 1965. The Palaeogene Beds of S.E. England. *Proceedings of the Geologists' Association*, Vol. 76, 151–173.

EHLERS, J, and LINKE, G. In press. The origin of deep buried channels of Elsterian age in north-west Germany. *Journal of Quaternary Science.*

ELLISON, R A, and LAKE, R D. 1986. Geology of the country around Braintree. *Memoir of the British Geological Survey.* Sheet 223 (England and Wales).

ESSEX RIVER AUTHORITY. 1971. First survey of water resources and demands. (Chelmsford: Essex River Authority.)

FUNNELL, B M. 1987. Late Pliocene and early Pleistocene stages of East Anglia and the adjacent North Sea. *Quaternary Newsletter*, No. 52, 1–11.

GALLOIS, R W, and MORTER, A A. 1982. The stratigraphy of the Gault of East Anglia. *Proceedings of the Geologists' Association,* Vol. 93, 351–368.

HARMER, F W. 1900. The Pliocene deposits of the east of England. II, The Crag of Essex (Waltonian) and its relation to that of Norfolk and Suffolk. *Quarterly Journal of the Geological Society of London*, Vol. 56, 705–738.

HESTER, S W. 1965. Stratigraphy and palaeogeography of the Woolwich and Reading Beds. *Bulletin of the Geological Survey of Great Britain*, No. 23, 117–123.

HEY, R W. 1965. Highly quartzose pebble gravels in the London Basin. *Proceedings of the Geologists' Association*, Vol. 76, 403–420.

— 1980. Equivalents of the Westland Green Gravels in Essex and East Anglia. *Proceedings of the Geologists' Association*, Vol. 91, 279–280.

HOPSON, P M. 1979. The sand and gravel resources of the country north of Harlow, Essex. Description of 1:25 000 resource sheet TL41. *Mineral Assessment Report of the Institute of Geological Sciences*, No. 46.

— 1981. The sand and gravel resources of the country around Stansted Mountfitchet, Essex. Description of 1:25 000 resource sheet TL52. *Mineral Assessment Report of the Institute of Geological Sciences*, No. 104.

KING, C. 1970. The biostratigraphy of the London Clay in the London district. *Tertiary Times*, Vol. 1, 13–15.

— 1981. The stratigraphy of the London Clay and associated deposits. *Tertiary Research Special Paper*, No. 6. 158 pp.

KNOX, R W O'B. In press. Thanetian and early Ypresian chronostratigraphy in south-east England. *Tertiary Research.*

— and ELLISON, R A. 1979. A Lower Eocene ash sequence in SE England. *Journal of the Geological Society of London,* Vol. 136, 251–253.

— and HARLAND, R. 1979. Stratigraphical relationships of the early Palaeogene ash-series of NW Europe. *Journal of the Geological Society of London,* Vol. 136, 463–470.

— — and KING, C. 1983. Dinoflagellate cyst analysis of the basal London Clay of southern England. *Newsletter, Stratigraphy,* Vol. 12, 71–74.

LAKE, R D, ELLISON, R A, HENSON, M R, and CONWAY, B W. 1986. The geology of the country around Southend-on-Sea. *Memoir of the British Geological Survey,* Sheets 258 and 259 (England and Wales).

LEE CONSERVANCY CATCHMENT BOARD. 1974. Report of a Survey of the Water Resources and Demands made by the Board in accordance with Section 14 of the Water Resources Act 1963. (Cheshunt: Lee Conservancy Catchment Board.)

MARKS, R J. 1982. The sand and gravel resources of the country between Hatfield Heath and Great Waltham, Essex. Description of 1:25 000 sheets TL51/61. *Mineral Assessment Report of the Institute of Geological Sciences,* No. 52.

— 1982. The sand and gravel resources of the country north-east of Thaxted, Essex. Description of 1:25 000 sheet TL63. *Mineral Assessment Report of the Institute of Geological Sciences,* No. 133.

MATHERS, S J, and ZALASIEWICZ, J A. 1988. The Red Crag and Norwich Crag formations of southern East Anglia. *Proceedings of the Geologists' Association,* Vol. 99, 261–278.

MILLWARD, D, ELLISON, R A, LAKE, R D, and MOORLOCK, B S P. 1987. The geology of the country around Epping. *Memoir of the British Geological Survey,* Sheet 240 (England and Wales).

MITCHELL, G F, PENNY, L F, SHOTTON, F W, and WEST, R G. 1973. A correlation of Quaternary deposits in the British Isles. *Special Report of the Geological Society of London,* No. 4. 99 pp.

MONKHOUSE, R A, and RICHARDS, H J. 1982. Groundwater Resources of the United Kingdom. *Atlas of groundwater resources. European Economic Community,* 252 pp. (Hanover: Th. Schäfer.)

MORTON, A C. 1982. The provenance and diagenesis of Palaeogene sandstones of south-east England as indicated by heavy mineral analysis. *Proceedings of the Geologists' Association,* Vol. 93, 263–274.

MURRAY, K H. 1986. Correlation of electrical resistivity marker bands in the Cenomanian and Turonian Chalk from the London Basin to east Yorkshire. *Report of the British Geological Survey,* Vol. 17, No. 8.

PERRIN, R M S, DAVIES, H, and FYSH, M D. 1973. Lithology of the Chalky Boulder Clay. *Nature, London,* Vol. 245, 101–104.

PERRIN, R M S, ROSE, J, and DAVIES, H. 1979. The distribution, variation and origins of pre-Devensian tills in eastern England. *Philosophical Transactions of the Royal Society of London,* Series B, Vol. 287, 535–570.

PRESTWICH, J. 1847. On the main points of structure and probable age of the Bagshot Sands, etc. *Quarterly Journal of the Geological Society of London,* Vol. 3, 378–409.

— 1852. On the structure of the strata between the London Clay and the Chalk, etc. Part iii. The Thanet Sands. *Quarterly Journal of the Geological Society of London,* Vol. 8, 253–264.

— 1890. On the relation of the Westleton Beds and Pebbly Sands of Suffolk, to those of Norfolk and on their extension inland. *Quarterly Journal of the Geological Society of London,* Vol. 46, 84–181.

ROSE, J (editor). 1983. *The diversion of the Thames. Field guide for annual meeting, Hoddesdon, 1983.* (Quaternary Research Association.)

— ALLEN, P, and HEY, R W. 1976. Middle Pleistocene stratigraphy in southern East Anglia. *Nature, London,* Vol. 263, 492–494.

— — 1977. Middle Pleistocene stratigraphy in south-east Suffolk. *Quarterly Journal of the Geological Society of London,* Vol. 133, 83–103.

SAYER, A R, and HARVEY, B I. 1965. Records of wells in the area of New Series One Inch (Geological) Great Dunmow (222) and Braintree (223) sheets. *Water Supply Paper of the Geological Survey of Great Britain.*

SOLOMON, J D. 1932. On the heavy mineral assemblages of the Great Chalky Boulder Clay. *Geological Magazine,* Vol. 69, 314–320.

— 1935. The Westleton Series of East Anglia: its age, distribution and relations. *Quarterly Journal of the Geological Society of London,* Vol. 91, 216–238.

SPARKS, B W, and WEST, R G. 1965. The relief and drift deposits. 18–40 in *The Cambridge region.* STEERS, J A (editor). (British Association for the Advancement of Science.)

— — WILLIAMS, R B G, and RANSOM, M. 1969. Hoxnian interglacial deposits near Hatfield, Herts. *Proceedings of the Geologists' Association,* Vol. 80, 243–267.

STAMP, L D. 1921. On cycles of sedimentation in the Eocene strata of the Anglo-Franco-Belgian Basin. *Geological Magazine,* Vol. 58, 108–114, 146–157, 194–200.

THAMES CONSERVANCY. 1969. Water Resources Act 1963: report of survey. (Reading: Thames Conservancy.)

THOMAS, C W. 1982. The sand and gravel resources of the country around Great Dunmow, Essex. Description of 1:25 000 resource sheet TL62. *Mineral Assessment Report of the Institute of Geological Sciences,* No. 109.

TURNER, C. 1970. The Middle Pleistocene deposits at Marks Tey, Essex. *Philosophical Transactions of the Royal Society of London,* Series B, Vol. 257, 373–437.

WARD, G R. 1978. London Clay fossils from the M11 Motorway, Essex. *Tertiary Research,* Vol. 2, 17–21.

WHITAKER, W. 1866. On the Lower London Tertiaries of Kent. *Quarterly Journal of the Geological Society of London,* Vol. 22, 404–435.

— 1921. The water supply of Buckinghamshire and of Hertfordshire from underground sources. *Memoir of the Geological Survey of England and Wales.*

— PENNING, W H, DALTON, W H, and BENNETT, F J. 1878. The geology of the NW part of Essex and the NE part of Herts with parts of Cambridgeshire and Suffolk. *Memoir of the Geological Survey of Great Britain,* Sheet 47, (England and Wales).

— and THRESH, J C. 1916. The water supply of Essex from underground sources. *Memoir of the Geological Survey of Great Britain.*

WHITE, H J O. 1932. The geology of the country around Saffron Walden. *Memoir of the Geological Survey of Great Britain.*

WILSON, D. 1983. Periglacial structures. 160–162 in *The diversion of the Thames.* ROSE, J (editor). (Quaternary Research Association.)

WOODLAND, A W. 1945. Water supply from underground sources of Cambridge–Ipswich district: Part VII—Well Catalogue for New Series One-Inch Sheets 221 (Hitchin) and 222 (Great Dunmow). *Wartime Pamphlet, Geological Survey of Great Britain*, No. 20 (VII). 83pp.

— 1970. The buried tunnel-valleys of East Anglia. *Proceedings of the Yorkshire Geological Society*, Vol. 37, 521–578.

WORSSAM, B C and TAYLOR, J H. 1969. The geology of the country around Cambridge. *Memoir of the Geological Survey of Great Britain*, Sheet 188 (England and Wales).

WRIGLEY, A. 1924. Faunal divisions of the London Clay illustrated by some exposures near London. *Proceedings of the Geologists' Association*, Vol. 35, 245–259.

— 1940. The faunal succession in the London Clay illustrated in some new exposures near London. *Proceedings of the Geologists' Association*, Vol. 51, 230–245.

APPENDIX 1

Abstracts of selected borehole logs referred to in the text

Great Bardfield Borehole   (TL 63 SE 1) NGR 6690 3091
Drilled in 1976 by BGS   Logged by M Heath and R A Ellison
Surface level +68.2 m

	Thickness m	Depth m
Kesgrave Sands and Gravels	1.8	1.8
London Clay	10.3	12.1
Woolwich and Reading Beds	c.14.0	c.26.1
Thanet Beds	c. 8.1	34.2

See also Figure 3

Little Chishill Borehole   (TL 43 NE 1) NGR 4528 3637
Drilled in 1964 by Superior Oil (UK) Ltd
Datum level of Kelly Bar +135.0 m

	Thickness m	Depth m
Drift	6.1	6.1
Upper Chalk	54.9	61.0
Middle Chalk	109.7	170.7
Lower Chalk	45.7	216.4
Gault	61.6	278.0
Devonian	10.6	288.6

See also reinterpreted Chalk sequence on p.4, based mainly on geophysical logs

Saffron Walden Fire Station Well   (TL 53 NW 72)
NGR 5386 3840
Drilled in 1836
Surface level c. +52 m
Interpretative log: 'chalk-marl' was recorded below 84.4 m

	Thickness m	Depth m
Drift	3.0	3.0
Upper Chalk	16.5	19.5
Middle Chalk	64.9	84.4
Lower Chalk	?54.9	?139.3
Gault	?42.7	?182.0
Other formations	?126.8	?308.8

Ware Well   (TL 31 SE 57) NGR 3531 1398
Drilled in 1879
Surface level +33.6 m

	Thickness m	Depth m
Drift	5.2	5.2
Upper Chalk	?55.8	?61.0
Middle Chalk	?69.2	130.2
Lower Chalk	49.6	179.8
Upper Greensand	12.2	192.0
Gault	50.8	247.8
Wenlock Beds, dipping 41°S	5.6	253.4

APPENDIX 2

1:10 000 maps

The following is a list of 1:10 000 geological maps included in the area of 1:50 000 geological sheet 222, with the names of the surveyors and the dates of survey of each map. The surveyors were R A Ellison, M J Heath, A Horton, R D Lake, P I Manning, S R Mills, D Millward, B S P Moorlock, G Richardson, D Wilson and J A Zalasiewicz.

Manuscript copies of the maps are deposited for public reference in the library of the British Geological Survey at Keyworth. Uncoloured dyeline copies of these maps are available for purchase from the British Geological Survey, Keyworth, Nottingham NG12 5GG.

TL41NW	Widford	BSPM	1975
TL41NE	Spellbrook	MJH	1976
TL42NW	Furneux Pelham	GR	1984-85
TL42NE	Manuden	RDL	1980
TL42SW	Little Hadham	GR	1983
TL42SE	Bishop's Stortford	RDL, JAZ	1980
TL43NW	Great Chishill	AH	1983
TL43NE	Elmdon	AH, PIM	1983
TL43SW	Brent Pelham	AH, PIM	1983
TL43SE	Clavering	PIM	1982-83
TL51NW	Little Hallingbury	MJH	1976
TL51NE	Hatfield Broad Oak	MJH	1976
TL52NW	Elsenham	RDL, DM	1980
TL52NE	Broxted	DM	1979
TL52SW	Birchanger	DM	1979
TL52SE	Takeley	DM	1979
TL53NW	Saffron Walden	JAZ	1980
TL53NE	Wimbish	JAZ	1980
TL53SW	Newport	DW	1980
TL53SE	Debden	DW	1980
TL61NW	High Roding	RAE	1976
TL61NE	Ford End	RDL	1969
TL62NW	Lindsell	DW	1979
TL62NE	Bran End	DW, SRM	1975, 1979
TL62SW	Great Dunmow	RDL	1979
TL62SE	Felsted	DW, SRM	1975, 1979
TL63NW	Great Sampford	JAZ	1979
TL63NE	Cornish Hall End	JAZ, BSPM	1976, 1979
TL63SW	Thaxted	JAZ	1979
TL63SE	Finchingfield	JAZ, BSPM	1976, 1979

APPENDIX 3

Open-file reports

The open-file reports listed below are detailed accounts of the geology of selected 1:10 000 sheets which form part of the Great Dunmow (222) 1:50 000 sheet. Copies of the reports may be ordered from the British Geological Survey, Keyworth, Nottingham.

Sheet	Area	Author, Year
TL42NW, and SW	Furneux Pelham and Little Hadham	G Richardson, 1986
TL42NE and SE	Bishop's Stortford and Manuden	R D Lake, 1981
TL43NW, NE, SW and SE	Great Chishill, Elmdon, Chrishall, Anstey, Nuthamstead and Clavering	A Horton, 1988
TL52NE, SW and SE	Broxted, Takeley and Stansted Mountfitchet	D Millward, 1980
TL53SW and TL52NW	Newport and Elsenham	D Wilson and R D Lake, 1981
TL53NW and NE	Saffron Walden and Sewards End	J A Zalasiewicz, 1981
TL53SE	Thaxted and Debden	D Wilson, 1981
TL62NW, NE, SW and SE	Great Dunmow	D Wilson and R D Lake, 1980
TL63NW, NE, SW and SE	Great Sampford, Cornish Hall End, Thaxted and Great Bardfield	J A Zalasiewicz, 1981

APPENDIX 4

Geological Survey photographs

Copies of the photographs are deposited for reference in the British Geological Survey library at the Keyworth Office. Prints and slides may be purchased. The photographs listed below were taken by Mr C J Jeffery and are available in colour and black and white. They belong to Series A.

13538	Contorted glacial silts, Ugley Sand Quarry [5194 2795]
13539–41	Glacial Sand and Gravel, Ugley Sand Quarry [517 280]
13542	Degraded Glacial Sand and Gravel, near Wade's Hall [513 280]
13543	Cryoturbated Glacial Sand and Gravel, near Wade's Hall [513 280]
13544	Weathering of Glacial Sand and Gravel, Walpole Farm [5135 2595]
13545	Solution pipes in Upper Chalk, Alsa Street [515 255]
13546	Kesgrave Sands and Gravels, Elsenham Sand Pit [550 265]
13547–8	Boulder clay resting on Kesgrave Sands and Gravels, Elsenham Sand Pit [550 258; 550 265]
13549	Kesgrave Sands and Gravels, Elsenham Sand Pit [550 265]
13550	Reclamation of gravel pit site, Elsenham Sand Pit [547 264]
13551	Solifluction materials, Pinchpools Farm [490 280]
13552	Fractured Upper Chalk, Pinchpools chalk pit [492 276]
13553	Disturbed Upper Chalk, Pinchpools chalk pit [492 276]
13554–9	Kesgrave Sands and Gravels, Cowlands Farm, Stebbing [670 233]
13560	Chelmer valley, near Great Easton
13561	Chalky boulder clay, Hollow Lane Gravel Pit, near Widdington [5317 3066]
13562	Boulder clay resting on Kesgrave Sands and Gravels, Hollow Lane Gravel Pit [5319 3072]
13563	Face in Upper Chalk, Newport chalk pit [526 332]
13564–8	Solution pipes in Upper Chalk, Newport [527 331]
13569–70	Glacial Sand and Gravel, Quendon [5172 3100]
13571	The upper Cam valley, from near Hollow Lane, Widdington [5283 3112]
13572	Glacial Silt, Widdington [5241 3203]
13573–5	Kesgrave Sands and Gravels, Great Sampford sand pit [641 361]
13576	Fill overlying Glacial Sand and Gravel, Great Sampford sand pit [641 361]

INDEX

Page numbers in italics refer to figures and tables.

Albury Hall   21
  chalk quarry   5
  gravel pit   21
Alisocysta margarita   12
alluvium   *16*, 30, 40–41
Alsa Street   8
  gravel pit   5
Alterbidinium minor   12
Anglian glacial channels   28–39
Anglian Glaciation   17, 18, 40
Anglian ice-sheet   2, 18, *27*
Anglian Stage   *16*, 17, *34*
Anglian Water Authority   42, *43*
Apectodinium paniculatum   12
A. summissum   12
aquiclude   42, 43
aquifer   42, 43
Armigers gravel pit   24, 42
ash beds   2, 7
Ash valley   1, 5, 42
  Quaternary   3, 19–21, 40
Audley End   1, 2

Barham Arctic Structure Soil   *16*, 18
Barham Loess   *16*
Barham Sands and Gravels   *16*
Beds   13–15 (Gault)   4
Beestonian Stage   *16*, 17
Birchanger   12, 22–24
  Interchange (M11)   13
Bishop's Stortford   2, 11
  Quaternary   *16*, 17, 18, 38–39
    buried channel deposits   30, 32
    river terrace deposits   40
  brick-making   11, 42
  Cricketfield Lane   11
  gravel pit, Hazelend   39
  Hockerill Street   39
Bonhunt   3, 30, 40
  Water   30, 36, 40
boreholes (named)
  Bishop's Stortford
    By-pass   39, 40
    Railway Hotel   32
  Bluegate Farm   15
  Bonhunt   36
  Chrishall   5
  Dunston Common   33
  Elsenham, Memorial Well   12
  Great Bardfield   8, 9, 10, 15, 47
  Great Sampford   42
  Hadham Park   11
  Little Chishill   4–5, 47
  M11 motorway   30, 36, 37
  Much Hadham Pumping Station   10
  Newport   30
  Quendon   30
  Saffron Walden   4, 5, 47
  Shaftenhoe End   5
  Stansted Mountfitchet, Limekiln Lane   40
  Stebbing, Tollesburies Farm   15
  Strethall   5
  Ware   4, 47
  Wendens Ambo   19
  Wickham Hall   11
Bottom Bed (Tertiary)   2
Bottom Bed facies (Woolwich and Reading Beds)   8, 43
boulder clay   1, 10, 18, 22, 32, 35, 39, 43
  chalky   36, 39
  decalcified   25
  see also till
Bozen Green   5
Braintree district   8, 14
Brakey Ley Wood   41
Bramertonian Stage   *16*, 18
Brand's Hill   35
brickmaking   42
brickpit/yard
  Bishop's Stortford   11
  Hadham Ford   10
  Tilekiln Green   24
Bromley   21
Bullhead Bed   8
  see Thanet Beds
buried channel   2, *16*, 18, 28, *34*
  see also channel-fill; glacial channel
Butcher's Pasture   41

calcareous tufa   *16*, 41
calcium-bicarbonate type of water   43
  carbonate   18, 41
  -sodium-bicarbonate type of water   43
calcrete   24
cambering   18
Cam valley   1, 2, 4, 42
  Quaternary   3, *16*, 18, 19, 21–22, 30, 32–38, 40, 41
Cam–Stort buried channel   28–39
  see also Stort valley
Canfield gravel pit   15, 42
Chaldean's Farm   21
Chalk   4–6, 8, 21, 22, 30, 36, 41
  as aquifer   42, 43
  relationship with buried channels   33
  see Lower Chalk; Middle Chalk; Upper Chalk
chalk   *6*, 21
  breccia   5, 30, 36, 39, 40, 41
  erratics/pebbles   21, 22, 24, 25, 35, 36, 38; *see also* chalk, reworked
  reworked
    Glacial Sand and Gravel   18, 21, 39
    Glacial Silt   37, 38
    Head   40, 41
Chalk Marl   5
Chalk Rock   4
channel-fill
  lithology   28
  Glacial Sand and Gravel   18, 24
  till   22, 28
  Cam–Stort buried channel   28–39
  structures   19, 25, 38
  see also buried channel

Chara marl   39, 40
Chelmer valley   1, 42
  Quaternary   3, 24, 40, 41
Chelmsford area   28, 33
Chillesford Sand Member   15, *16*, 18, 22
Chiloguembelina wilcoxensis   10
Churchend Farm   10
Clitherocytheridea sp.1   10
Colchester   28
Cowlands Farm gravel pit, Stebbing
  frontispiece, 15, 18, 24, *26*, 42
Crag deposits   2, 14–15, *16*, 17, 18, 19, 22
  as aquifer   42
  reworked   2, 15
Cretaceous   2, 4–6, *15*
  junction with Tertiary   7, 12
  junction with Quaternary   21, 30, 36, 38, 39, 40, 41
  see Chalk
Cromerian Stage   *16*, 17
cryostatic injection features   5
cryoturbation   23, 24, 35, 40
Cytheretta aff. *scrobiculoplicata*   10
Cytheropteron brimptoni   10

Danbury–Tiptree ridge   28
Debden Water   18, 30, 40
  gravel pit   36
Deflandrea oebisfeldensis   12
Devensian Stage   3, *16*, 28, 40
Devonian   2, 4, 47
diapiric structures   21
drainage   1, 2
  Quaternary   28, 33, 40, 42
dropstones   19, 37
Dunmowpark   41
Durrel's Wood   12

Elsenham   10, 12, 14
  Quaternary   *16*, 18, 22–24, 38–39
    buried channel deposits   30
  Cross   22
  Memorial Well   12
  railway cutting   22
  sand and gravel pit   15, 22, *23*, 42
Eocene Series   7, 8–13
  junction with Quaternary   41
Epistominella cf. *vitrea*   10
Epping district   10
Essex River Authority   42

Farnham Green   21
Felsted   18
Flandrian Stage   3, *16*, 40–41
flow till   24, 32
Foxdells   32, 39
Frambury Lane, Newport   36

Gault   2, 4, 42, 47
Gipping valley   28
glacial channel   28–39, 41
Glacial Sand and Gravel   2, 12, *16*, 18, *20*, 28, 30, 36, 39, 42
  lithology   18, 23, 24, 25, 35, 36, 37, 38, 39

Glacial Silt   2, 16, 19, *20,* 30, 33, 36
  lithology   35–38
Glandulina sp. 1   10
G. sp. 2   10
Glauconitic Marl   4, 5
Globulina ineaqualis   10
Glomospira? sp.   10
Grailands   11
Great Bardfield   14
  borehole   *9,* 47
    Tertiary   8, 10
    Plio-Pleistocene   15
Great Dunmow   2, 8, 14
  Quaternary   40, 41
Great Easton   18
Great Hormead   8
Great Sampford   14, 15
  Quaternary   16, 18, 41
  sand pit   18, 25, *27*
Gyroidina sp.   10

Hadham Ford   21
  brickyard   10
  see Little Hadham; Much Hadham
Hadham Hall   10
Hadham Park   11
Haplophragmoides sp.   10
Hatfield Heath   40
Hazelend sand and gravel pit, Bishop's
    Stortford   39, 42
Head   3, 10, 22, *16,* 30, 33, 40
  lithology   21, 22, 35, 36, 37, 38, 39, 41
Head Gravel   5, 19, 20, 40
Hollow Lane gravel/sand pit, Widdington
    12, 15, 18, 22, 30, 42
Howells   30, 37
Hoxnian Stage/interglacial   3, *16, 34,*
    40–41
hydrogeology   42–43
hyperacanthum Zone   12
Hystrichokolpoma mentitum   12

interglacial periods   2, 3
  see Hoxnian Stage/interglacial
intraformational ice-wedges   18

Jock Farm   30
Joshua's Bridge   41

Kesgrave Sands and Gravels
    frontispiece, 2, 5, 16, 17–18, 19, 21,
    22, *23,* 24, 25, 28, 33, 42, 47
  as aquifer   43
  lithology   9, 20, 21, 22, 23, 24–25,
    27
  reworked   5, 21, 22, 23, *27,* 30

Latchmore Bank   40
Late Glacial zones III–IV   40
Lee Conservancy Catchment Board   42
    Mountfitchet   40
Littlebury   30
  Nursery Lodge 33–34
Little Hadham   10, 40
  quarry   5
  see also Hadham Ford; Much Hadham
London Jock Wood   30, 38

London Clay   7, 8–13, 24, 47
  basal beds   2, 11, 12
  lithology   9, 23
Lower Chalk   4, 5, 42, 47
Lower Greensand   4
Lower London Tertiary Group   7, 8, *9*

M11 motorway   8, 10, 12, 13, 30, 36, 37
Maggots End   10
Marks Tey interglacial deposits   28
Matching Green   40
Melbourn Rock   4
mere   3, *16*
  deposits   40
Micraster coranguinum   *4*
  Zone   5
Middle Chalk   4, 5, 35, 47
  as aquifer   42
Morrice Green   40
Much Hadham   8
  see also Hadham Ford; Little Hadham
Mytiloides labiatus   4

NP9–NP10 zonal boundary   7, 8
Neville Hill Wood   35, 41
Newport   1, 4
  Quaternary channel deposits   30
  Frambury Lane   36
  Grammar School   35
  gravel pit/quarry   5, *6,* 35
  Pond   37, 40
Nonion cf. *graniferum*   10
N. cf. *laeve*   10
Nonionella sp.   10
North Hall   22, 30, 40, 42
Nursery Lodge, Littlebury   33–34
Nuthampstead   1

Palaeocene Series   7, 8
  see Thanet Beds; Woolwich and Reading
    Beds
Palaeozoic   2, 4
Pant valley   1, 42
  Quaternary   3, 25, *27,* 40, 41
Parsonage Farms   22
peat   40, 41
Pinchpools chalk pit   5, *6*
Plio-Pleistocene   14–15
Protelphidium cf. *hofkeri*   10
P. sp. 3   10
proto-Thames river   2
Pseudohastigerina wilcoxensis   10

Quaternary   3, 16–27, 28–39, 40–41
Quendon   3, 21
  channel deposits   30
  Park   30
  sand pit   21
  Want   40

'Reading Beds'   10
Reading type/facies   8, 43; *see* Woolwich
  and Reading Beds
Red Crag   14, 22, 42
Ringers Farm   21
River Terrace Deposits   3, *16,* 32, 40, 41
Roding valley   1, 42

rubified horizon   *frontispiece,* 17, 18, 19,
    24, 25

Shortgrove   35
  Hall   41
sodium-bicarbonate-chloride type of
    water   43
sol lessivé 16–19
solution features   5, *6,* 8, 21, 22, 41
Standon Lodge   10
Stansted   22
  Castle chalk pit   5, 12
Stansted Mountfitchet   2, 8, 12
  Limekiln Lane borehole   40
Start Hill   23, 24
Stebbing   14, 15, 18
  Brook valley   24, 25
  Cowlands Farm gravel pit
    frontispiece, 15, 18, 24, *26,* 42
  Tollesburies Farm borehole   15
Sternotaxis planus   4
Stort valley   1, 2, 42
  Quaternary   3, 16, 18, 19, 30, *31,* 32,
    33, 39, 40, 41
  see Cam–Stort buried channel
Sudbury district   10, 28

Thames Conservancy   42
Thames Water Authority   42, 43
Thanet Beds   2, 7, 8, 43, 47
  fauna   10
  lithology   9
Thanetian Stage   *8,* 12
Thaxted   2, 14
Tilekiln Green brickpit   24
Till   1, 10, 16, 17, 18, 19, 21, 22, 23,
    28, 30, 32, 33, 35, 37, 38, 40, 41
  as aquifer   43
Top Rock   4
Totternhoe Stone   5

Ugley   38
Ugley Green   8
  Quaternary   18, 19
  buried channel fill   30, 32
  sand pit   12, *19,* 30, 38
Upper Chalk   1, 4, 5, 10, 11, 12, 21, 22,
    39, 41, 47
  as aquifer   42
Upper Devonian   4
Upper Gault   2, 4
Upper Greensand   4, 47
Upper Jurassic   4

Valley Farm Rubified *Sol Lessivé*   16
volcanogenic material   8, 10, 24, *27*

Wendens Ambo   35, 41
  borehole   19
Wenlock   4, 47
Westland Green Gravels   16, 17
Wicken Bonhunt   40
Widdington   8, 12, 15
  Quaternary   16, 18, 40
  Hollow Lane sand and gravel pit   12,
    15, 18, 22, 30, 42
Widdington Sands   17

Wolstonian Stage *16*
Woolwich and Reading Beds 2, 7, 8, 21, 22, 47
 as aquifer 43
 economic use 42
 lithology *9*, 10, 11, 12, 13

Ypresian Stage 8

BRITISH GEOLOGICAL SURVEY

Keyworth, Nottingham NG12 5GG
(06077) 6111

Murchison House, West Mains Road,
Edinburgh EH9 3LA   031-667 1000

London Information Office, Natural History Museum
Earth Galleries, Exhibition Road, London SW7 2DE
071-589 4090

The full range of Survey publications is available through the Sales Desks at Keyworth, Murchison House, Edinburgh, and at the BGS London Information Office in the Natural History Museum, Earth Galleries. The adjacent bookshop stocks the more popular books for sale over the counter. Most BGS books and reports are listed in HMSO's Sectional List 45, and can be bought from HMSO and through HMSO agents and retailers. Maps are listed in the BGS Map Catalogue and the Ordnance Survey's Trade Catalogue, and can be bought from Ordnance Survey agents as well as from BGS.

The British Geological Survey carries out the geological survey of Great Britain and Northern Ireland (the latter as an agency service for the government of Northern Ireland), and of the surrounding continental shelf, as well as its basic research projects. It also undertakes programmes of British technical aid in geology in developing countries as arranged by the Overseas Development Administration.

The British Geological Survey is a component body of the Natural Environment Research Council.

Maps and diagrams in this book use topography based on Ordnance Survey mapping

HMSO

HMSO publications are available from:

HMSO Publications Centre
(Mail and telephone orders)
PO Box 276, London SW8 5DT
Telephone orders 071-873 9090
General enquiries 071-873 0011
Queueing system in operation for both numbers

HMSO Bookshops
49 High Holborn, London WC1V 6HB
  071-873 0011 (Counter service only)
258 Broad Street, Birmingham B1 2HE
  021-643 3740
Southey House, 33 Wine Street, Bristol BS1 2BQ
  (0272) 264306
9 Princess Street, Manchester M60 8AS
  061-834 7201
80 Chichester Street, Belfast BT1 4JY
  (0232) 238451
71 Lothian Road, Edinburgh EH3 9AZ
  031-228 4181

HMSO's Accredited Agents
(see Yellow Pages)

And through good booksellers